Soil Science Simplified

Fourth Edition

Soil Science Simplified

Fourth Edition

Milo I. Harpstead
Thomas J. Sauer
William F. Bennett

Illustrated by Mary C. Bratz

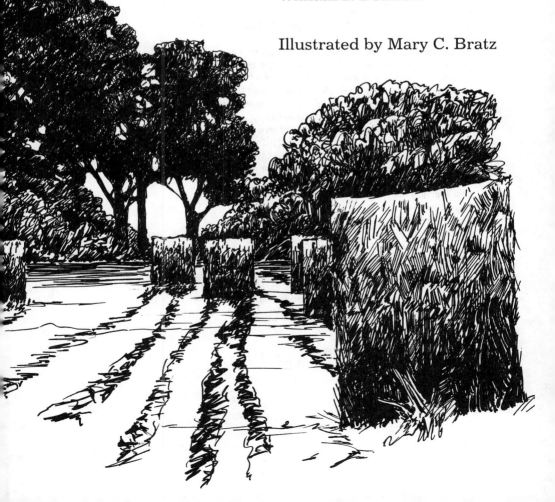

Milo I. Harpstead, retired, was a Professor of Soil
Science, University of Wisconsin–Stevens Point.

Thomas J. Sauer is a Soil Scientist with the U.S.
Department of Agriculture, Agricultural Research
Service, National Soil Tilth Laboratory, Ames, Iowa.

William F. Bennett, retired, was a Professor of
Agronomy–Soils, Texas Tech University, Lubbock.

Blackwell Publishing Professional
2121 State Avenue, Ames, Iowa 50014

Orders: 1-800-862-6657
Office: 1-515-292-0140
Fax: 1-515-292-3348
Web site: www.blackwellprofessional.com

Printed on acid-free paper in the United States of America

First edition, 1980
Second edition, 1988
Third edition, 1997
Fourth edition, 2001

Library of Congress Cataloging-in-Publication Data

Harpstead, Milo I.
     Soil science simplified / Milo I. Harpstead, Thomas J.
Sauer, William F. Bennett; illustrated by Mary C Bratz. —
4th ed.
        p. cm.
     ISBN 0-8138-2942-9 (alk. paper)
     1. Soil science. I. Sauer, Thomas J. II. Bennett, William F.
III. Title.
     S591 .H28 2001
     631.4—dc21   2001003279

The last digit is the print number: 9 8 7 6 5 4 3

CONTENTS

PREFACE

oil Science Simplified explains soil science in an easily understandable manner. Students, professionals, and nonprofessionals alike will gain an accurate working knowledge of the many aspects of soil science and be able to apply the information to their endeavors. The book is a proven and successful textbook and works well as assigned reading for university students in the natural sciences and earth sciences. Agricultural science courses taught at the high school or post high school level also can use this edition as a textbook.

At one time, soil science was largely directed toward agriculture. Farming remains at the forefront of food production and is, more than ever, concerned with soils; but the properties of soils affect everyone who works with soils. Horticulturists, foresters, landscape architects and home gardeners frequently seek an in-depth understanding of soils. Homebuilders must be aware of soil-based regulations, particularly for waste disposal. Engineers need to understand how the physical and chemical properties of soils react to human installations and manipulations. Environmentalists and people in related areas find a working knowledge of soils useful. And everybody who works with the land in any way needs to know how to take full advantage of the information in a soil survey report.

This fourth edition of *Soil Science Simplified* expands and updates each topic. New approaches to content have been incorporated to increase the reader's ease in understanding explanations. Some chapters have been reorganized, some have been combined, and all contain current information. The illustrations demonstrate the principles described in the text and enhance comprehension.

The authors are experienced university professors of soil science and researchers who grew up on farms and gained a basic, hands-on appreciation of the importance of applied soil science. This book represents their many years of experience and the desire to engender respect for soils.

Soil Science Simplified

Fourth Edition

CHAPTER 1

The Soil around Us

We depend on the soil in many ways. Since the birth of the soil conservation movement in the 1930s, there has been an increased interest in protecting the soil. The environmental awareness message of the past several decades has focused attention on soil as a fundamental part of the ecosystem to be preserved. In spite of this, there is little public understanding of soil's complexity. The main purpose of this book is to explain the basic principles of soil science in a practical way.

Careful observers may see soil exposed in roadbanks or excavations and they may notice that the soil does not look the same in all locations (Fig. 1.1). Sometimes the differences are apparent in the few inches of surface that the farmers plow, but greater variations can usually be seen by looking at a cross section of the top 3 or 4 feet (0.9 or 1.2 m) of soil. The quality of vegetative growth depends on the makeup of the soil layers, which are called *horizons*. Similarly, roads and structures may fail if they are constructed on unsuitable soils. Special care must be taken to overcome soil limitations for specific engineering uses. Satisfactory dis-

1.1. Roadbanks can reveal the complexity of the soil.

posal of human waste and livestock manure is becoming an increasingly great concern as more communities experience contaminated water supplies.

Poor yields of agricultural crops and poor growth of trees may result from a mismatching of crops and soils. This mismatching may happen because the landowner has not examined the soil horizons or understood their limitations. Soil scientists study the factors necessary for proper soil management.

What Is Soil?

The traditional meaning of *soil* is that it is the natural medium for the growth of land plants. In 1975 the Soil Conservation Service offered a more inclusive definition in the book *Keys to Soil Taxonomy*: "Soil is the collection of natural bodies on the earth's surface, in places modified or even made by man of earthy materials, containing living matter and supporting or capable of supporting plants out-of-doors." The 1999 edition of the same book presented a new definition of soil to take into account the soils of Antarctica, where the climate is too harsh to support higher plants. It is long and involved and is not presented here because the authors believe the earlier definition is adequate.

Most soil is a loose mass of fragmented and chemically weathered rock with an admixture of humus, which is partially decomposed organic matter. In the wet areas of humid regions, plant residue may accumulate several feet thick to form a peat soil, but in dry regions, soil organic matter may be low throughout the landscape. Soil is very diverse over the face of the earth, but if properly managed, it serves us well.

The Nature and Uses of Soil

Soil is indirectly a source of food. If there were no soil, the continents would be wastelands of barren rock. It is fortunate that over most of the land area of the earth soil covers bedrock to a considerable depth. In the porous soil, seeds germinate and roots grow as they obtain water and nutrients. In this way, crops of the fields and forests produce food and fiber.

Soil provides physical support and shelter. Soil gives mechanical support for plant roots so that even tall trees stand for decades against strong winds. Soil also physically supports structures such as buildings, sidewalks, streets, and highways. Sometimes this support is imperfect so that buildings crack and pavement breaks up due to the instability of the underlying soil. Some abandoned Roman roads have been buried by soil carried upward by ants, earthworms, and other creatures. To the millions of humans who walk or jog every day on bare-soil footpaths, the soil gives welcome support and softens the impact on their feet.

In intertropical regions, millions of people live comfortably in places built chiefly from locally excavated soil. Such earthen houses are common in West Africa (Fig. 1.2). A compound earthen dwelling for an extended family may be quite an impressive structure. The adobe houses of the southwestern United

1.2.   Earthen houses are common in West Africa.

States and the pioneer sod houses of the prairies are other examples of earthen houses. Modern earth-sheltered homes are shown with pride by the owners and builders. For maximum insulation, they feature an earthen embankment covering all but the side exposed to the sun.

Soil performs complex water management. Water in the form of rain, dew, fog, irrigation, or snowmelt that falls on or runs over a body of soil is divided by the soil into surface and subsurface water. The surface water may evaporate or may run off to provide excess water to streams. The water that soaks into the soil to become subsurface water may be taken up by plant roots and transpired from the leaves back into the atmosphere or, if there is enough of it, it may percolate down to become part of the groundwater reservoir from which rivers and springs are fed. Many organic and inorganic pollutants in wastewater are absorbed as it passes through the soil, giving us pure groundwater. However, if potential pollutants that are very soluble are added to the soil, they may be carried into the groundwater to our detriment.

Soil is an air-storage facility. Plant roots and billions of other organisms living in the soil need oxygen. The pore system in soil provides access to air, which is pushed into and drawn out of the soil by changes in barometric pressure, by turbulent wind, by the flushing action of rainwater, and simply by diffusion. Some plants, including rice, have the capacity to conduct oxygen into waterlogged soil. Soil air contains considerable amounts of carbon dioxide, which some bacteria can use as a source of carbon for their protoplasm.

Soil is even useful as a mineral supplement. In some impoverished African countries, selected types of soil have been used as a special food supplement. Specifically, pregnant women and their babies have benefited from the mother's ingestion of soil from termite mounds that are enriched with calcium. By using this natural resource, these women have bypassed agricultural products for adequate calcium.

1.3. Living organisms sooner or later become a part of the soil once again.

Soil accepts back that which came from it. When plants die, it is not long before organisms that cause decomposition go to work on them; even huge logs on the forest floor soon disappear (Fig. 1.3). Animals that live on plants as well as other forms of life also return to the soil when they die. Our society produces vast amounts of waste of every size, shape, and description; much of it is buried in landfills with the hope that it will return to the soil.

Soil is beautiful; it is an aesthetic resource. People may become fond of their native soil, whether it is black and brown as is usual in the temperate climatic regions or red and yellow closer to the equator. There is a rainbow of various hues of soil under our feet. Changes in both soil and vegetation through the seasons are observed with great interest. Some soils form wide cracks in dry seasons and swell when the rains return. Frost action may create delicate little ice pillars that lift the surface of the ground in winter. The smell of freshly tilled soil seems good to farmers and gardeners as they plant their crops in the spring with high expectations for an abundant autumn harvest. Some people love their native soil so much that even today they still perform the ancient ritual of kneeling to kiss it when they return home.

How Big Is an Acre? A Hectare?

There are many situations in agriculture where the application of soil amendments are assigned in units of pounds per acre, and these numbers are quite

1.4. An acre is 208.7 feet on a side; a hectare is 100 meters on a side.

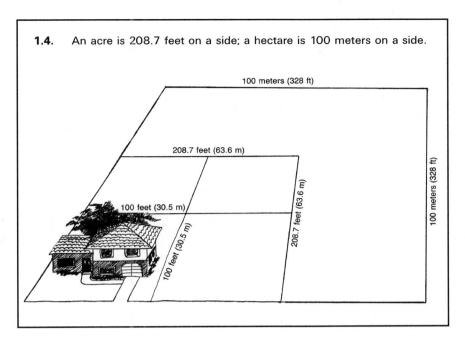

close to the corresponding metric units of kilograms per hectare. Likewise, the loss of soil by erosion is rated in tons per acre, which closely correspond to metric tons per hectare.

As the world's population becomes more urbanized, units for measuring land in either the English or metric system become vague to most people. Many house lots in cities have an area of about 10,000 square feet, whereas an acre has 43,560 square feet. Figure 1.4 puts some of these units in perspective.

CHAPTER 2

Soil Formation

Excluding the organic component, the solid portion of soil is primarily a product of rock weathering. To understand the weathering of rocks, it is essential to consider how rocks are formed and how that formation influences their mineralogy. The minerals in rocks strongly influence the composition of the soil derived from them. The other factors in nature that influence the specific properties of soil in a given location also will be discussed in this chapter.

One of the basic rules of nature is that nothing remains the same over long periods of time. Astronomers tell us that even stars such as our sun have a finite lifespan. They coalesce from cosmic dust, form into shining solar bodies, finally expend their energy, collapse, and return to cosmic dust. The secrets of these processes have only recently started to be revealed by the Hubble telescope. On earth, the alteration of rocks from one form to another is much more easily understood because we can study specimens of rocks and relate them to their position in the earth's crust.

The term *rock* refers to a material within a specified range of mineralogical composition that is of appreciable extent in the crust of the earth. Some of the most common rocks are granite, basalt, sandstone, and limestone. Rounded pieces of rock—so common in glaciated regions—are boulders, stones, cobbles, and gravels in descending order of size.

The Rock Cycle

To understand the formation of soil, consider first the rocks from which the mineral particles in the soil were derived. As the earth cooled, the molten magma crystallized into igneous rocks. As long as there has been water on the earth, flowing water has been eroding rocks and the fine particles produced have become sediments, which may solidify into sedimentary rocks. Under conditions of extreme heat and pressure, both igneous and sedimentary rocks may be modified and at least partially recrystallized into metamorphic rocks.

2.1. The rock cycle shows how heat and pressure, melting and erosion cause rocks to change in form through geologic time.

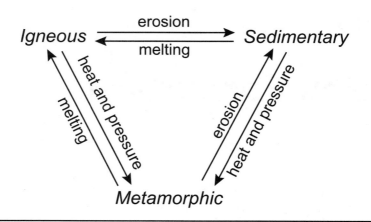

The shifting of continents causes landmasses to slide over and bury other landmasses to the extent that the buried ones may become molten again. Where this occurs there is evidence of great tectonic activity in the form of earthquakes, volcanoes, faults, and related phenomena. Therefore, over geologic time the rocks of the earth are cycled from one form to another (Fig. 2.1). Rocks are the evidence for these actions in the past, and the same processes continue today.

Composition of the Earth's Crust

Chemists recognize a few over 100 elements that make up everything tangible on earth. Of these, the eight listed below are the most abundant elements in the earth's crust. The others are no less important but are present in much smaller quantities.

Element	Ion
Oxygen	O^{--}
Silicon	Si^{++++}
Aluminum	Al^{+++}
Iron	Fe^{++}, Fe^{+++}
Calcium	Ca^{++}
Magnesium	Mg^{++}
Potassium	K^{+}
Sodium	Na^{+}

Silicate Minerals

If molten magma from within the earth cools very rapidly, these elements solidify randomly into a glass such as obsidian, a material commonly used in jewelry. If the cooling is slower, the elements will assemble themselves into crystalline silicate minerals. The slower the cooling is, the larger the crystals.

Silicates are minerals made up, in large measure, of combined silicon and oxygen. They are the most common minerals in rocks. When only silicon and oxygen ions are involved, they form a four-sided structure with oxygen ions at the points and a silicon ion in the center. It can be compared to a three-sided pyramid with the base being the fourth side. This is called a tetrahedron. If the O^{--} on each corner is shared with another tetrahedron, a very strong framework structure results. A mineral with this form is *quartz*, and it is so resistant that it is said to be nonweatherable.

Of the silicates, most are *aluminosilicates*; *feldspars* are the classic example. Feldspars also have a framework structure but from one-fourth to one-half of the Si^{++++} was replaced with Al^{+++} during the original crystallization of the feldspar. Since Al^{+++} has a lower positive charge than Si^{++++}, the unsatisfied negative bonds from the O^{--} are satisfied primarily by K^+ and Ca^{++} in the crystal. Feldspars are quite stable but are less resistant to weathering than is quartz. The weathering of feldspar accounts for much of the potassium and calcium found in the soil, the oceans, and sedimentary rocks.

Micas are the other main group of aluminosilicates. The tetrahedra are formed into layers that can be lifted, one from the other, like the pages of a book. When separated from the rocks, these small flat particles will glisten in the sun, especially if they settled out of flowing water and lie flat on the dried soil surface.

Most of the very dark colored minerals in rocks are *ferromagnesian silicates*. Instead of the framework silicate structure discussed above, these minerals have single, paired, or chained sets of tetrahedra that are bonded together by accessory ions, usually Fe^{++} and Mg^{++}, hence, the term ferromagnesian. It is by way of the accessory ions that weathering gains access to these minerals and the integrity of the mineral structure is destroyed. Two common groups of these dark minerals are the *amphiboles* and *pyroxenes*.

Igneous Rocks

Igneous rocks (Fig. 2.2), including granites and their metamorphic associates, make up the bedrock foundation of the continents. The minerals in them are crystalline in form and, if the magma cooled slowly far below the surface of the earth, the crystals are comparatively large. This is the case with granite. If the cooling of the magma took place more rapidly, the crystals are small, such as in rhyolite. Granite and rhyolite may be identical in mineralogical composition and are characterized by having abundant quartz due to the high silica content of the magma. In a parallel manner, magma lower in silica may solidify into very dark colored gabbro or basalt, depending on the rate of cooling.

Crystalline igneous rock lies just below the unconsolidated surface material on about one-quarter of the earth's land area. Elsewhere it is more deeply buried. It is quarried for building stones and monuments. Pink and light-colored granite is popular. It outcrops dramatically in the Black Hills of South Dakota at Mount Rushmore,

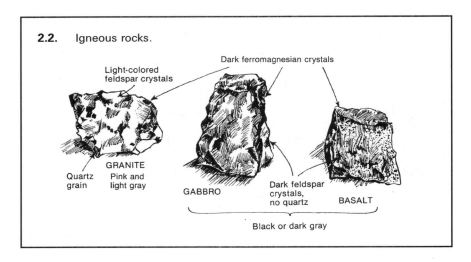

2.2. Igneous rocks.

Dark ferromagnesian crystals

Light-colored feldspar crystals

GRANITE

Quartz grain

Pink and light gray

GABBRO

Dark feldspar crystals, no quartz

BASALT

Black or dark gray

where the heads of four U.S. presidents have been carved. Gabbro can be polished into a beautiful building stone and is sometimes called black granite. A well-known example of it is the Vietnam Memorial in Washington, D.C. Blackish and finely crystallized basalt is well known because of extensive volcanic activity on earth.

Sedimentary Rocks

Sedimentary rocks (Fig. 2.3) are the bedrock for about three-quarters of the land area of the earth. These rocks were deposited as loose layers of sediment on the bottoms and edges of ancient seas. Sand, primarily quartz grains, was deposited near the shores, gray siliceous mud farther out, and limy, whitish mud from fossil shells in the deep water. These layers gradually hardened into rock to become sandstone, shale, and limestone, respectively. As the land was slowly uplifted and the seas receded, sedimentary rock covered most of the continents.

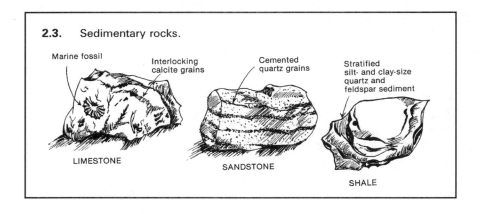

2.3. Sedimentary rocks.

Marine fossil

Interlocking calcite grains

Cemented quartz grains

Stratified silt- and clay-size quartz and feldspar sediment

LIMESTONE

SANDSTONE

SHALE

Metamorphic Rocks

Rocks can be altered by heat and pressure within the earth. The metamorphic rocks that result may have been any of the igneous or sedimentary rocks. Granite is commonly metamorphosed into gneiss, a beautifully banded rock wherein like minerals became concentrated due to similar viscosity in the shifting magma. Sandstone is cemented by silica from solution to become quartzite, which is the most resistant rock that is widespread on the earth. Shale is converted into slate and limestone into marble by heat and pressure.

Processes of Rock Weathering

When living organisms such as plants die, they are rotted by saprophytic microorganisms. In a similar manner, naturally occurring physical and chemical forces cause rocks to be weathered into *saprolite*. Collectively, saprolite is called the *regolith* of the earth, which is composed of the loose mineral materials above solid bedrock. The effects of rock weathering can be observed by splitting a stone that has been exposed on or near the ground surface for a long time (Fig. 2.4). During the weathering process, the altered rock material may accumulate in place over the solid rock or it may slide, be washed, or be blown to other sites. Soil formation begins soon after loose rock material is stabilized.

Physical Weathering

Physical weathering of rocks is their breakdown into progressively smaller pieces with no change in molecular arrangement within the minerals. Any of the forces that transport solid particles causes them to wear. Sand on a beach that is rolled by each incoming wave is a familiar example. Strong winds pick up sand and blast it against objects that soon show the effects of abrasion. Tree roots penetrate cracks in rocks and as the roots grow they cause the cracks to expand and eventually break the rock. In temperate regions, water enters cracks in rocks, freezes, and can cause the surface of the rock to peel off like the rind of an orange, which is called exfoliation. Glaciers were the ultimate in physical weathering as they broke loose massive boulders and moved them great distances with a grinding action. Hills were lowered, valleys were filled, and there was a general leveling effect except at the glacier's edge, where terminal moraines were built up.

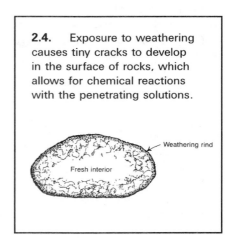

2.4. Exposure to weathering causes tiny cracks to develop in the surface of rocks, which allows for chemical reactions with the penetrating solutions.

Weathering rind

Fresh interior

No matter the extent of physical weathering, it does not directly cause significant release of ions from the minerals for the benefit of plants.

Chemical Weathering

Chemical weathering, as the term implies, results from chemical reactions that alter the molecular composition of minerals. These chemical forces react with the surface of minerals. If physical weathering did not greatly increase surface area by breaking down rock into smaller pieces, chemical weathering would progress much more slowly.

Hydrolysis. Hydrolysis has been called public enemy number 1 for minerals. It takes place when hydrogen ions (H^+) in water replace metallic ions in minerals. All water is slightly ionized, so hydrolysis is pervasive. Rain absorbs carbon dioxide (CO_2) as it falls, resulting in a relatively weak carbonic acid (H_2CO_3), which greatly increases the reaction of hydrolysis. However, the ancient statues from the Greek and Roman empires did not show much degradation until smoke from modern industry resulted in sulfuric and nitric acid in the precipitation. Over millions of years, most of the acid in the soil resulted from the respiration of CO_2 by living organisms. Plant roots also release H^+ during nutrient uptake.

In one simplified example of hydrolysis, potassium feldspar reacts with water to yield silicic acid and potassium hydroxide. The silicic acid is the building block of clay. In the reaction, the primary mineral is destroyed, clay is formed, and potassium (K^+) is released into the soil for use by plants. If precipitation is sufficient to leach away the base (KOH), the land will become more acid and the sea will become more basic.

$$KAlSi_3O_8 + HOH \rightarrow HAlSi_3O_8 + KOH$$

Oxidation. Oxidation takes place when certain multivalent ions lose an electron (a negative charge) to become more positive. A common element in rocks capable of two valence states is iron. Just as a wrench left in the rain will rust, so also iron-bearing minerals in rocks become oxidized. The equation for the reaction is

$$Fe^{++} \rightarrow Fe^{+++} + e^-$$
$$4\,Fe^{+++} + 3\,O_2 \rightarrow 2\,Fe_2O_3$$

Hydration. By itself, oxidation would not be extremely disruptive to the mineral, but, in nature, it is followed by hydration:

$$Fe_2O_3 + nH_2O \rightarrow Fe_2O_3 \cdot nH_2O$$

In this reaction, n water molecules attach themselves to an iron oxide molecule. This results in considerable expansion, which greatly disrupts the mineral structure of the rocks and causes them to crumble. For this reason, when digging in the subsoil in a humid region it is common to encounter stones that disintegrate if struck by a spade. Manganese is another element in minerals that can exist in multivalent ionic states, but it is much less abundant than iron. Salt may also hydrate with similar results.

Reduction. Reduction, being the opposite of oxidation, reflects a gain of electrons in multivalent ions. It is not disruptive to most bedrock, but it does have a marked influence on soil where oxygen has been depleted by microorganisms in wet places. Under reducing conditions, iron and manganese may be dissolved and removed from the system or translocated to regions with free oxygen. Here they precipitate as nodules, concretions, or various types of layers and coatings.

Solution. Water comes as close as anything to a universal solvent. However, it is only capable of dissolving large quantities of soluble salts that were precipitated from solution at some earlier time. The calcium carbonate in limestone came from the shells of sea creatures. It is an example of a salt that is slowly soluble in pure water, but water enriched by carbonic acid, due to biological activity, reacts with limestone. This dissolution is evidenced by massive caves where the rock was dissolved by biologically acidified water that seeped down from the soil.

Factors of Soil Formation

Soil scientists think of soils as natural bodies that have length, breadth, and depth. Each soil body occupies a portion of the landscape. This means that soils are more than simply the product of rock weathering; they are components of the landscape (Fig. 2.5), just as are rivers, forests, marshes, and prairies. Thousands of years have been required to make our present-day soils. Five factors of soil formation have been identified (Fig. 2.6). They are (1) parent material, (2) climate, (3) living organisms, (4) topography, and (5) time.

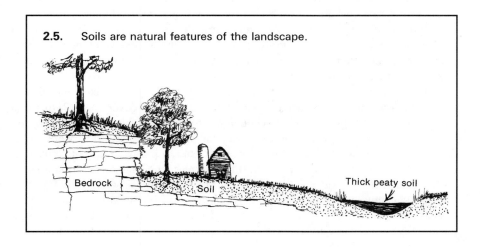

2.5.   Soils are natural features of the landscape.

Bedrock

Soil

Thick peaty soil

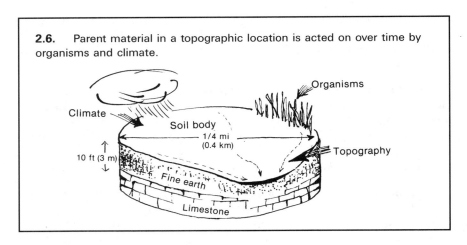

2.6. Parent material in a topographic location is acted on over time by organisms and climate.

Soil Parent Material

Parent material of mineral soils is the weathered rock that was slowly broken up at a site or was transported there by natural agents. It can be grouped into (1) crystalline rocks, such as granite and gneiss, (2) sedimentary rocks, such as sandstone and limestone, and (3) geologically recent cover deposits, such as alluvium and glacial till.

Soils that have formed from granite contain a full range of particle sizes, from gravel and sand to the finest clay. Since quartz grains (somewhat like bits of glass) in granite are very resistant to weathering, they become the gritty sand in the soil. The less-resistant minerals in rock—such as feldspar (a word meaning field crystal) and dark minerals rich in iron and magnesium (ferromagnesian minerals), including black mica—are altered by weathering into fine clay particles.

Black and dark gray crystalline rocks include gabbro (coarse grained) and basalt (fine grained). Because these rocks contain no quartz, soils formed from gabbro and basalt are not sandy but are clayey, sticky, commonly quite red, and rather fertile.

Soils from sandstone are sandy; those from shale are silty or clayey. Soils from limestone consist largely of insoluble shaley materials that were included as gray mud in the otherwise more weatherable rock mass. Therefore, soils from limestone commonly are clayey.

Recent cover deposits are blankets of geologically young sediments that overlie the types of bedrock just discussed. They include (1) cover sand, (2) loess, (3) volcanic ash, (4) glacial drift, (5) alluvium, (6) and colluvium (Fig. 2.7).

Cover sands are most common in arid and subhumid areas. Most were initially deposited by water when massive expanses of sandstone were being eroded over a long period. Wind action may shift these cover sands into dune formations, which are then referred to as *eolian* deposits. The Sand Hills of western Nebraska are a good example of eolian deposits. When viewed from an airplane, they are seen as an expanse of crescent-shaped dunes. They are droughty and not very productive for crops or livestock.

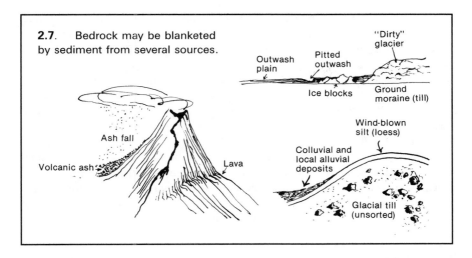

2.7. Bedrock may be blanketed by sediment from several sources.

Loess is a wind-transported deposit that consists mainly of silt that was derived from the flood plains of rivers that drained the meltwater from glaciers. These silts have a rich supply of plant nutrient-bearing minerals, and their size is such that they hold a significant quantity of water for crops. Extensive areas of fertile agricultural soils can be found in loess deposits in such places as China, the Mississippi–Missouri valley, and the Danube valley in Europe.

Volcanic ash is widespread in Hawaii, Oregon, and Washington in the United States and in Central America, Japan, Indonesia, and many other mountainous areas. The mineralogy of volcanic ash is variable, but most of it develops into high-quality soil for crop production.

Glacial deposits, often with a covering of loess, are parent materials of soils in much of the corn belt in North America and the wheat belt of Eurasia. They were left by glaciers (and their meltwaters) that advanced and retreated repeatedly between 1 million and 10,000 years ago. Glaciers carried a lot of rock debris collected by a grinding action on the terrain over which they passed and thus were made of "dirty" ice. An unsorted mixture (till) of stones, sand, silt, and clay was deposited in broad blankets and ridges called moraines. Glacial till is sometimes stony enough to inhibit cultivation, but its fresh supply of minerals provides an abundance of many plant nutrients. Rapidly flowing meltwaters left behind extensive sheets of sand and gravel, called outwash, that tend to be droughty for crops. Where huge ice blocks, which melted later, were surrounded by glacial drift (till and outwash), large pits or potholes were formed. Many lakes once existed near the glaciers. Today the ancient lake bottoms are almost level farmlands with rich silty and clayey soils.

Alluvium is sediment that was deposited by rivers and streams in valleys throughout the world. Centuries of erosion have created fertile areas of alluvial soils: the Bangkok plain, the Mekong delta, the Mississippi delta, and the vast alluvial plains of China. About one-third of the human population is supported on these fertile flood plains, which are rich in topsoil materials brought down from the uplands. Although flooding is a major hazard to humans, buildings, and crops, it is a major agent in depositing soil materials. Alluvial soils are finely layered (stratified) to great depths. Each layer may represent the deposit of a single flood. These soils show marked changes horizontally, from somewhat sandy near river-banks on natural levees and alluvial fans to clayey and even peaty in remote swampy areas. Older soils with distinct subsoil layers may be found on natural terraces, or "high bottoms," that now stand above the rest of the valley floor but were subject to flooding at one time (Fig. 2.8).

Colluvium, a gravity-transported deposit at the base of foothills or moun-tains, moved from above to its present location. Often, as in the case of mudflows, it was in a somewhat fluid state at the time of transport. These deposits are ex-tremely variable in composition but are not geographically extensive. Colluvium includes talus, which consists of chunks of broken rock at the foot of a mountain.

Climate

Every place on earth has climate that can be described based on its many components. The two components that most strongly influence soil formation are precipitation and temperature.

Each of the soil-forming factors interacts with the others, and this is evident with climate. It strongly influences the rate at which rocks are weathered into a loose regolith. It controls the supply of water for physical weathering and deter-mines breakup by freezing and thawing. Climatic change led to the advancement and retreat of glaciers and the resulting glacial till.

It is the effect of climate on chemical weathering that has the greatest in-fluence on the weathering of rocks. Precipitation provides the water essential for chemical weathering processes and may be sufficient to carry away soluble

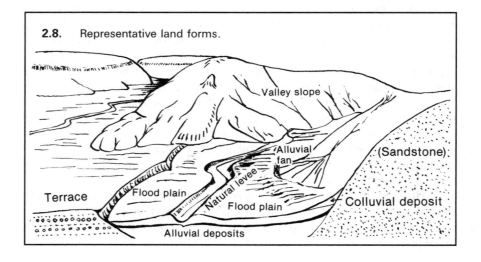

2.8.   Representative land forms.

Valley slope

Alluvial fan

(Sandstone)

Terrace   Flood plain   Natural levee   Flood plain   Colluvial deposit

Alluvial deposits

products, thereby allowing the reaction to continue. Without water, there can be no hydrolysis or hydration. Even oxidation-reduction may be dependent on the quantity of dissolved oxygen. The solution of minerals in certain rocks is dependent on rainfall unless they are adjacent to a body of water.

Temperature has a marked influence on the rate of soil formation. Perhaps the most obvious effect is that which occurs in the temperate zone, where essentially no chemical weathering takes place while the ground is frozen. There is a well-established rule in chemistry that for every 10-degree rise in temperature on the Celsius scale, the rate of chemical reactions increases by a factor of two to three. For example, the soils of the warmer southern part of the United States are more highly weathered than those in the cooler northern states even where glaciers were not a factor.

The combined influence of precipitation and temperature is probably as important as either one of them individually. If the temperature is cool, water does not evaporate fast, so the effectiveness of the precipitation is high. On the other hand, some warm areas receive quite a lot of precipitation, but due to rapid evaporation, they have the properties of a much drier climate. As an example, St. Paul, Minnesota, and San Antonio, Texas, each receive about 28 inches of precipitation annually, but because of the cool Minnesota temperature, the soil there is normally moist, whereas in the San Antonio area, the soil is usually dry. This effect also is reflected in the native vegetation, which is hardwood forest in the St. Paul area and drought-tolerant vegetation in the prairies of South Texas.

Living Organisms

The influence of all the organisms, plants and animals (both large and small) is the biotic factor of soil formation. Chapter 4 is devoted to soil biology, but in this section the ways that living organisms are involved in soil development are discussed.

In any particular climatic region, the amount of humus in the soil is a direct result of how much and what type of plant residue has been incorporated into it. Thus, if vegetation is sparse, the soil will be low in humus and less fertile. Grasses have a fibrous root system that quite thoroughly invades the tiny pores of the soil so that as the roots live, die, and decay over thousands of years, the soil becomes well supplied with humus. Tree roots are much larger, but because they do not invade the pores of the topsoil as completely as those of grasses, the humus content of soils under forests is usually lower.

Most of the trees in the world's forests can be divided into two groups: the hardwoods with broad leaves and the softwoods (conifers) with needles. Chemical analyses of broad leaves and needles show that needles are usually more acid because they contain fewer base-forming elements such as calcium and magnesium. Grasses contain even more bases than either broad leaves or needles (Fig. 2.9). Therefore, soils formed under conifer forests tend to be the most acid and least well buffered (for example, against acid rain).

2.9. Grass leaves are normally highest in bases, broad leaves of trees are intermediate, and conifer needles are the lowest.

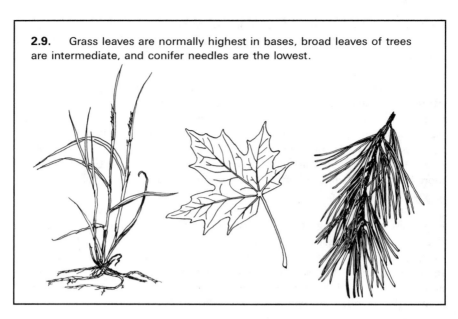

Grassland regions have the most fertile soil for agriculture, but most of them are subject to extended dry periods. Pioneers tended to select the hardwood forests as places to settle because the soils were quite good, and they needed the forest products for their livelihood.

Topography

The lay of the land—its levelness or hilliness—is called *topography*. Topography influences the formation of soil primarily in two ways. (1) Erosion carries topsoil from the higher positions, particularly the sideslopes of hills, and deposits it in the valleys. This results in relatively thicker, more-fertile soils in the valleys. (2) Water drains from the uplands to the valleys and if the excess is removed in a timely manner, vegetation is more abundant there. The abundant plant life, which does not decompose as rapidly in moist valleys as on the drier uplands, also contributes to the formation of deep, dark-colored, fertile soils. As a result, much of the world's population relies on crops grown in valleys for their food.

Climatic conditions modify the effects of topography on soil development. In the subhumid and drier climates, the soils are well drained in all positions in the landscape, but they differ in thickness by their long history of erosion or deposition. In the humid regions with a rolling landscape, soils may be thin and excessively drained on the hills and thick with poor drainage in the valleys. Broad, nearly level topographic positions typically have deeply developed soils even if they lie high above the drainage-ways. In the humid regions, these areas will show the effects of excess moisture unless the parent material is coarse textured so it will allow rapid internal drainage. In semiarid regions, broad level uplands typically have deep, dark colored soils formed under grassland vegetation.

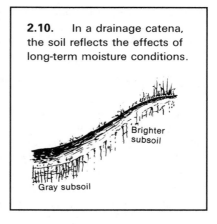

2.10. In a drainage catena, the soil reflects the effects of long-term moisture conditions.

Topography is a strong indicator of soil characteristics within a particular region. In 1935 an English earth scientist named Milne, working in East Africa, noticed the sequential nature of soils from the top of one hill, down through the valley, up the next hill, and down again repeatedly. Being a scholar in classic languages, Milne knew the Greek term for the arc formed by a chain suspended between two posts. From this, he derived the term *catena*, meaning a sequence of soils differing from each other due to their topographic position. In a two-dimensional sense, he saw each soil as a link in the chain. A ***catena*** is a *toposequence of soils* that may differ from each other in a variety of ways, such as composition and drainage. The drainage catena relationship of a humid region illustrated in Figure 2.10 is horizontally compressed.

Time

Time is typically discussed as the last of the five soil-forming factors. It is a consideration of how long the other factors have been influencing soil formation. The effects of time can best be seen in equatorial regions, where the extremes in age are well expressed. Geologically young areas typically have an irregular topography, and they are comparatively more fertile because young parent materials usually contain an abundance of weatherable minerals that slowly release plant nutrients as they weather. Geologically old surfaces, on the other hand, have long since lost most of their weatherable minerals. Their fertility is found primarily in the organic matter, which is subject to rapid depletion under cultivation. Since prehistoric times, farmers in the tropics have been attracted to rugged landscapes because of the success of growing better crops there. Similar comparisons of soil fertility could be made between geologically young regions such as the northern Rocky Mountains and old, highly weathered portions of the Piedmont Plateau in the southeastern United States.

In glaciated regions, which occur in much of the northern part of the United States, there is a relationship between the time since the last glacial advance, the irregularity of the landscape, and the degree of soil development as evidenced by the concentration of clay in the subsoil. Regions with more recent glacial till (<25,000 years) have many undrained depressions that may form lakes and potholes. Moderate to steep slopes are common, and the leaching of clay to the subsoil is moderate. In regions where the glacial till is much older (>50,000 years), more of the depressions have been filled and a complete drainage pattern has formed, so potholes are scarce. The slopes here are more gentle, and there is usually a much greater concentration of clay leached into the subsoil (Fig. 2.11).

Soil Horizon Development

During soil formation both parent materials and organic materials are altered and translocated so that layers called *soil horizons* develop. The layers usually can be recognized visually. A cross section of soil horizons, called a *soil profile*, is exposed when a pit or roadside is excavated. Two profiles are illustrated in Figure 2.12. One is typical of some of the subhumid grasslands and the other depicts the soil of humid hardwood forest regions.

Although the number and properties of these horizons vary widely, a rather typical soil profile in a humid region is discussed in this section. Dark humic materials commonly accumulate in the topsoil (the A horizon), followed by a leached zone (E horizon—from the word *eluvial* meaning washed out). The subsoil (B horizon) commonly has an accumulation of clay. The depth to the bottom of the B horizon is typically the depth to which there are abundant plant roots and biological activity. Certainly some roots may extend much deeper.

The portion of the soil profile that has been altered by the soil-forming factors is called the *solum* and is made up of the A, E, and B horizons. On the surface of the A horizon, there may be a layer of plant residue called an O horizon. Below the B horizon the underlying unconsolidated material is called the C horizon. If bedrock is within a few feet of the surface, it is called the R horizon. These symbols may be subdivided with small letters and numbers because of the diverse nature of soil. This system provides symbols used in making detailed soil profile descriptions. The symbols are a type of shorthand used by soil scientists, and they reveal much about the soil properties. The principal soil horizons can be categorized into diagnostic horizons, which are discussed in Chapter 11.

Leaching of plant nutrients such as potassium and calcium takes place as water moves through the soil, but some nutrients are retained by the finely divided humus and clay materials. Plants take up these nutrients and transport them into

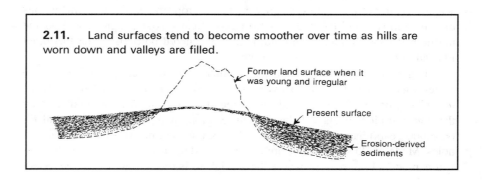

2.11.   Land surfaces tend to become smoother over time as hills are worn down and valleys are filled.

Former land surface when it was young and irregular

Present surface

Erosion-derived sediments

2.12. The profile on the left illustrates a soil from a subhumid grassland; the one on the right shows a soil from a humid hardwood forest region.

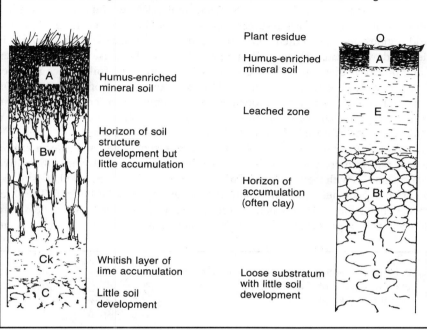

A — Humus-enriched mineral soil

Bw — Horizon of soil structure development but little accumulation

Ck — Whitish layer of lime accumulation

C — Little soil development

Plant residue

Humus-enriched mineral soil — O, A

Leached zone — E

Horizon of accumulation (often clay) — Bt

Loose substratum with little soil development — C

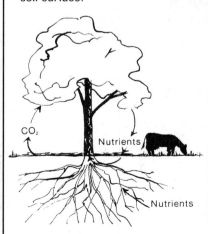

2.13. Biotic cycling helps to concentrate nutrients near the soil surface.

CO_2

Nutrients

Nutrients

their aboveground parts. The nutrients are returned to the soil as the seasons progress; thus, plants contribute to nutrient recycling. This biotic cycling helps to keep the soil from becoming infertile by frequent leaching (Fig. 2.13). Weathering is an ongoing process in the soil and to a lesser extent in the substratum below. As soil ages, it is likely to have a higher clay content because clay results from the physical and chemical breakdown of larger particles.

Let's Take a Trip

As we travel from one climatic region to another, there are distinct changes in the native vegetation, and if the farm fields have been plowed, there are differences in the appearance of the soil, even to the casual observer. If the soil is exposed to some depth, there are even more changes evident to those who examine the subsoil carefully.

If we take a trip in the United States from the deserts of the West to the humid woodlands of the East, a succession of soils could be seen (Fig. 2.14). In the arid regions, the tan-colored soil is only a little darker on the surface than it is deeper down because meager rainfall provides for only sparse vegetation. Even here, however, there are differences. Salts may whiten the soil surface in lower areas if water containing large amounts of salts evaporates off the surface. On very old geologic surfaces, carbonates may accumulate in the subsoil to form rocklike layers. Pebbles scattered on these ancient surfaces are likely to have a dark reddish-brown varnish from oxides of iron and manganese.

As our trip takes us into the central midwestern states, we enter a region where rain is more common during the growing season, and where the native prairie grasses with their abundant fibrous roots have made the topsoil thick, dark, and rich in plant nutrients. These soils do not have a leached E horizon. They are, in the main, the most productive soils in the United States. When fields are plowed, they appear almost black from the abundant humus, and if a road is cut

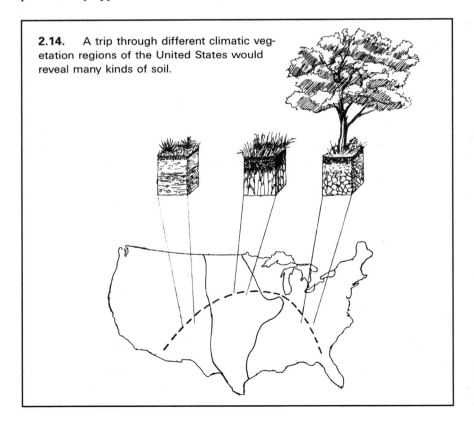

2.14. A trip through different climatic vegetation regions of the United States would reveal many kinds of soil.

through them, they show that the humus commonly extends 2 or more feet (61 cm) below the surface.

As the average rainfall and humidity increase toward the eastern one-half of the United States, forests replace the grasslands, and the soils are markedly different. When they are tilled, these soils have a grayish-brown appearance, which reflects their lower content of humus and the presence of a leached E horizon beneath a thin A horizon. The subsoil usually has a concentration of clay that shows up as a reddish-brown horizon in roadcuts or other exposures. Many of these soils are very productive, but they require more fertilizer and lime because leaching by greater rainfall has occurred.

If we swing south across the Ohio River, we find soils that are geologically much older, and soils in which the effects of weathering have been greater. Here the cultivated fields are quite red in most places as a result of iron from the minerals that have become oxidized. In these soils, the clay-enriched subsoil forms a much thicker zone, and their native fertility is low.

Whenever you have the opportunity to travel, be alert to the changes in the soil and try to relate them to the differences in the soil-forming factors discussed in this chapter.

CHAPTER 3

Soil Physical Properties

The soil properties that can be seen or felt are physical and are discussed in this chapter. Chemical properties cannot be seen or felt but can be detected with sophisticated scientific instruments. Chemical properties can be altered to our needs with soil amendments, but physical properties are much more permanent and difficult to change. Thus, physical properties should receive greater consideration in land-use planning.

Soil Phases

From a physical standpoint, soil is a three-phase system: solid, liquid, and gas. Each phase is equally essential for growth of plants. The solid phase is made up primarily of minerals along with a small amount of humus in most soils. This phase provides a source of nutrients and anchorage for plants and makes up approximately half of the soil volume. The liquid and gaseous phases are in the soil pores and occupy the other half. The proportion of each varies as the soil gains or loses moisture. Plants must be able to absorb water from the soil, and all except a few aquatic plants depend upon the soil pores for the oxygen that is essential for every cell in their roots. Figure 3.1 illustrates the approximate proportion of all three phases in a moist soil.

Soil Separates

A microscopic view of a section through a small piece of soil is shown in Figure 3.2. The white areas are pore spaces filled with various proportions of air or water, depending on the moisture conditions. The striped bodies are sand grains. Notice that each sand grain is coated with a film of tiny clay particles and

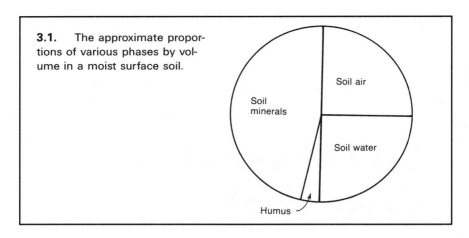

3.1. The approximate proportions of various phases by volume in a moist surface soil.

Soil air

Soil minerals

Soil water

Humus

that clay (represented by dots) forms connections between sand grains and silt particles (shaded with cross-hatching). The silt particles are smaller than sand but larger than clay. Particles of these three sizes together constitute the *fine earth* fraction of the soil.

If the diameter of medium-sized particles of clay, silt, and sand were expanded 1,000 times, the clay would have a diameter about the thickness of this page, the silt about 1 inch (2.5 cm), and the sand about 40 inches (1 m). *Coarse earth* is made up of gravel and stones that have a diameter greater than $\frac{1}{12}$ inches (2 mm), which is about the thickness of the lead of a pencil.

Sand

Sand ranges in diameter from 2 mm to 0.05 mm and is divided into five classes (Table 3.1). The smallest sand grains are only $\frac{1}{40}$ the size of the largest,

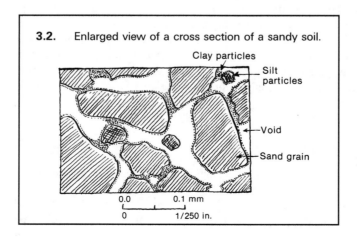

3.2. Enlarged view of a cross section of a sandy soil.

Clay particles

Silt particles

Void

Sand grain

0.0   0.1 mm

0   1/250 in.

and a wide range in their grittiness can be expected. If moist soil is rubbed between the fingers, grittiness caused by the sand grains can be felt.

Sand forms the framework of soil and gives it stability when in a mixture with finer particles. Pure

Table 3.1. Sizes of sand particles

Class	Diameter (mm)
Very coarse sand	2.0–1.0
Coarse sand	1.0–0.5
Medium sand	0.5–0.25
Fine sand	0.25–0.1
Very fine sand	0.1–0.05

sand, however, does not cling together, so it is easily eroded by water and wind. During erosion, sand is not suspended in the water or air but bounces along the surface and piles up where the velocity decreases. In the case of wind erosion, this causes sand to form into drifts like snow.

Quartz is usually the dominant mineral in sand because it is the most resistant to weathering of the common minerals in rocks; thus, its breakdown is extremely slow. Many other minerals are found in sand, depending on the rocks from which the sand was derived.

The shape of sand grains is more or less spherical. However, the angularity of sand grains is variable due to the degree to which the specific deposit was rolled around by flowing water.

Sand contributes very little to plant nutrition. The quartz in sand contributes no plant nutrients to the soil while the other minerals, such as feldspars, release their nutrients very slowly. Nevertheless, soils that have a lot of feldspar and other weatherable minerals in their sand fraction develop a comparatively higher state of fertility over the thousands of years of soil formation.

Silt

In many respects, silt is similar to sand except that it is smaller, having a diameter of 0.050 to 0.002 mm, which is too small to be seen with the naked eye. It is spherical and mineralogically similar to sand.

Silt is too fine to be gritty to the touch but imparts a smooth feel without stickiness. It is fine enough to be suspended in flowing water, but it drops out when the flow is reduced. This is the reason that harbors are said to become "silted in." If silt is disturbed by drifting sand, it can be picked up and carried great distances by strong winds; thus, silt constitutes the main part of the wind-deposited parent material, loess. This concept will be discussed further in Chapter 10.

Clay

This soil separate is for the most part much different, particularly in size and chemical composition, from sand and silt. Sand and silt are progressively finer and finer pieces of the original crystals in the parent rocks, while clay, on the other hand, is made up of secondary minerals that were formed by the drastic alteration of the original forms or by the recrystallization of the products of their weathering. Clay is so powdery fine that 1 g, which has a volume about equal to that of a pencil eraser, may have a total surface area equal to one-fifth of a football field (Fig. 3.3). This tremendous surface area results from the plate-like shape of the individual clay particles. The maximum diameter of a clay particle is 0.002 mm, and the finer colloidal clays are in the range of 0.0001 mm. They can only be viewed clearly with an electron microscope.

To illustrate some characteristics of clay, take a large ball of pie dough and roll it into a thin sheet with a rolling pin (Fig. 3.4). Pieces cut from the sheet could be stacked to make a model of a clay particle. The pile of thin sheets would have a much larger surface area, inside and out, than the original ball of dough. Similarly, each clay particle is actually a stack of many very small sheets. There are many kinds of clay, each with different internal arrangements of chemical elements that give them individual characteristics. The major groups of clays are discussed in more detail in Chapter 5.

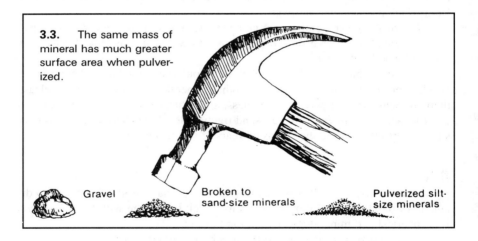

3.3. The same mass of mineral has much greater surface area when pulverized.

Gravel   Broken to sand-size minerals   Pulverized silt-size minerals

3.4. A layered clay crystal is similar in nature to a stack of thin sheets of dough.

External surfaces   Internal surfaces

Soil Texture

Soil texture is the degree of fineness or coarseness of the soil. It is an expression of the relative amounts or percentages of sand, silt, and clay. Rubbing soil between the thumb and forefinger is a simple field test that can be used to estimate soil texture. If moist soil feels slippery, but not really sticky, it is a silty soil. If it is very sticky and can be rubbed into a cohesive ribbon that extends from the fingers like a broad blade of grass, it is a clayey soil.

A soil with a significant amount of sand, silt, and clay is called a *loam*. Various kinds of loams are classified by feel according to the degree of grittiness, smoothness, and stickiness: sandy loam, silt loam, and clay loam. A simple loam without excessive amounts of any ingredient has about 20% clay, 40% silt, and 40% sand (Fig. 3.5). Compared to silt and sand, clay is so sticky that not much is required to give the soil a special texture.

A textural triangle (Fig. 3.6) can be used to show the domains of the various soil textures. The word *loam* applies to the central area in the lower part of the triangle. Equal amounts of sand, silt, and clay would be a clay loam. Using the textural triangle, determine the texture of a soil with 50% sand, 20% silt, and 30% clay. The texture is a sandy clay loam. Additions of humus to a soil (not shown in the triangle) modify soil behavior; sandy soils seem finer textured and clay soils seem coarser textured than they really are.

The texture of a soil does not indicate how it got that way. Did wind, water, or glacial ice drop the particles of sand, silt, and clay at a particular site? Such questions about the processes of soil landscape formation were discussed in Chapter 2.

Particle Size Analysis

To make an accurate measurement of the amounts of sand, silt, and clay in a soil, a sample of soil is placed in distilled water to which a detergent has

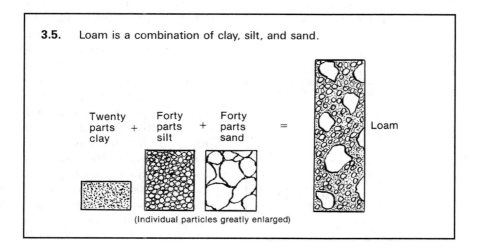

3.5. Loam is a combination of clay, silt, and sand.

Twenty parts clay + Forty parts silt + Forty parts sand = Loam

(Individual particles greatly enlarged)

3.6. A textural triangle shows the limits of sand, silt, and clay content of the various texture classes.

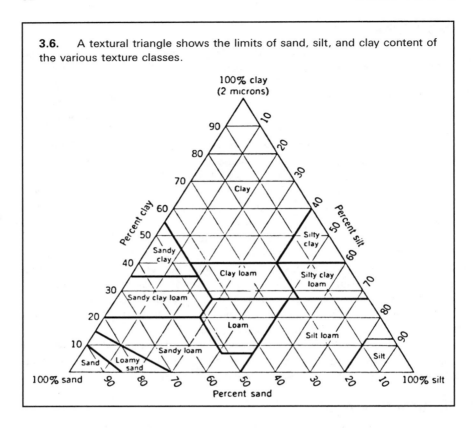

3.7. The amounts of different-sized soil particles can be measured.

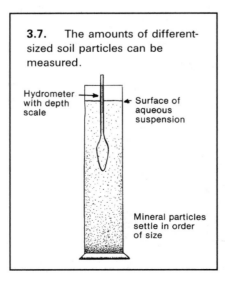

been added to separate the soil particles and then it is shaken (Fig.3.7). The mineral particles settle out at different rates; the sand falls to the bottom first, the silt next, and the clay settles out only after many hours or days. A bulblike float, called a hydrometer, is placed in the column of water. It floats highest when the soil is in suspension. The hydrometer sinks as the soil settles. The position of the water surface as seen against a scale on the side of the hydrometer indicates how to calculate the texture of the soil according to the time elapsed. Directions for making this kind of analysis of particle size in a soil sample may be obtained from agricultural laboratories and many soil science textbooks.

Porosity and Density

The percentage of pore space in soil is called *porosity*. As the parent material of soil becomes weathered, loosened, and mixed by a variety of forces, pore space develops, providing a place for air and water to be held. Both the amount of pore space and the size of the pores are important. Small pores retain water very well while, in large pores, water drains out and air moves in. Therefore, it is desirable to have both large and small pores in the soil.

How loose or tight a soil is defines its density. Density of soil, called *bulk density*, includes both the solid particles and the pore spaces among them. If a soil is compacted, the amount of pore space is reduced, and the weight of a given volume of soil is increased. The measure of density is a comparison to water, which has a density of 1 gram per cubic centimeter (1 g/cm³). The mineral grains in the soil have a density of about 2.6 g/cm³. The total volume of the soil is around 40% to 60% pore space, so by using a mean value of 50% for porosity, bulk density would be 1.3 g/cm³. This is one-half the density of the minerals in solid rock (Fig. 3.8). Density can be expressed in the English system, such as pounds per cubic foot, but it is customary to express density in metric units.

Some soils have naturally compacted layers (pans) that may have a high bulk density. Such densities restrict root penetration and water movement. In other cases, heavy tractors and machinery may cause serious compaction (Fig. 3.9), which limits plant growth. In recent years, there has been a shift toward the use of tillage equipment that properly loosens the soil, leaves some protective crop residue on the surface, and allows for fewer trips to be made over the field.

Composition of Soil Pores

Soil pores can be filled totally with air or water. If a medium-textured soil is moist but freely drained, the air and water content of its pores are probably about equal. Normally, soils that seem dry still contain some moisture and the relative humidity in the pores remains near 100%. The water in the pores is actually a soil solution because it contains the ions of dissolved salts.

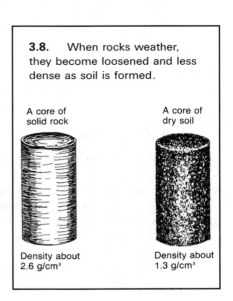

3.8. When rocks weather, they become loosened and less dense as soil is formed.

A core of solid rock

A core of dry soil

Density about 2.6 g/cm³

Density about 1.3 g/cm³

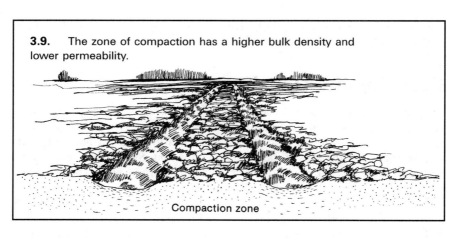

3.9.   The zone of compaction has a higher bulk density and lower permeability.

Compaction zone

Some are plant nutrients that may be absorbed by plant roots. The soil solution may also contain organic compounds such as humic and fulvic acids. Humic acid, for example, frequently gives the soil solution a brownish tinge. An abundance of dissolved humus may give alkali (sodic) soils a very dark brown color, but this condition is not widespread.

The earth's atmosphere is about 78% nitrogen (N_2), 20.9% oxygen (O_2), and 0.03% carbon dioxide (CO_2) with trace amounts of other gases. If the surface soil has free exchange, the soil air and the atmosphere will have about the same composition. However, when the plant roots and soil organisms are flourishing in the warm seasons, CO_2 is being respired by the living cells as oxygen is being absorbed. Nitrogen is essentially inert for all but a few specialized organisms, so its content remains unchanged. Oxygen and CO_2 are the main variables. It is common in the root zone for O_2 to drop to 10% and the CO_2 to rise to 10% without ill effects to the plants. Even 5% O_2 and 15% CO_2 may not be harmful, but crops vary in their tolerance to CO_2. When soil pores fill with water, the life-sustaining O_2 is soon depleted. Corn is very sensitive to this condition, but sorghum can withstand several days of flooding without permanent damage.

Soil Structure

Individual soil particles of sand, silt, and clay tend to become clustered into units of various shapes. This clustering is referred to as *soil structure*, which is defined as the arrangement of soil particles. The resulting structural units may be called aggregates or *peds*. Structural arrangements result from physical forces that cause some movement within the soil. Among these forces are shrinking and swelling from moisture variation, freezing and thawing, and the actions of expanding roots and of earthworms and other soil organisms. The weak cementing agents that hold aggregates together may be clay, organic compounds, oxides (particularly of iron), and various salts. Without soil structure, water would not readily enter the soil, roots would not develop well, and tuber growth would be restricted.

Soil structure units are classified according to shape as granular, platy, blocky, prismatic, and columnar structures (Fig. 3.10).

Granular structure is best recognized by farmers and gardeners who strive for a mellow soil. The more-or-less spherical clusters are called *aggregates*, and when soil is tilled it can be determined if it is well aggregated by the ease of working it. Some clay and a plentiful amount of organic matter are the keys to stable aggregates in the topsoil. The aggregates in sandy topsoil are usually rather porous, like breadcrumbs, and it is described as having a **crumb structure**. If there is a lack of stability, the aggregates will not withstand the forces of raindrops and tillage. If this happens, they break down and the fine particles close off the pores so that instead of moving into the soil, the rainwater runs over the surface with the potential for accelerated erosion.

Platy structure has a long horizontal and a short vertical axis. When this occurs in the subsoil, water penetration is restricted. For example, on-site waste disposal systems for rural homes are likely to fail if soil beneath the seepage bed has platy structure.

3.10. Soil structural units are classified according to shape.

Soil structural units (peds)	Soil structural types	Common occurrence
	Granular	In dark surface soil
	Platy	In pale subsurface soil
	Blocky	In clay-enriched subsoil, particularly in forested regions
	Prismatic	In subsoil, particularly in grassland regions
	Columnar	In sodium-affected subsoil of grassland regions

Blocky structure is the most common structure in the subsoil in humid regions that had forest as its native vegetation. The vertical and horizontal axes are about the same length. This gives a somewhat cubical form that allows good water percolation along the boundaries of the blocks. If there is plentiful clay in the soil, the edges of the blocks are likely to be angular. Structure is less well developed in sandy soils and edges of the blocks are rounded. This is known as **subangular blocky structure**.

Prismatic structure is best developed in the subsoils with a plentiful amount of clay in regions where the soil becomes periodically desiccated. These conditions are most common where prairie grasses were the native vegetation. The sides of the prisms act as an avenue for water movement.

Columnar structure is an undesirable variation of prismatic wherein the tops of the prisms are rounded and usually covered with gray soil particles. If the topsoil is cleared away, the tops of the columns look like the tops of baking powder biscuits. This happens when there is too much sodium in the soil. This condition is extremely restrictive to water percolation but, fortunately, it is usually localized in semi-arid regions.

Soil Consistence

A description of soil consistence gives an indication of how soil will react to mechanical manipulation at various moisture contents. The field measurements are made between the fingers, which give a good indication of how the soil will react to tillage, traffic, digging, or similar activity. When the soil is dry, it is described according to a fixed set of parameters as to its degree of hardness or softness. In the moist state, the degree of friability or firmness is used. When wet, it is ranked by its stickiness. The amount and type of clay is the single most important characteristic in determining soil consistence. For example, a clayey soil is likely to be very hard when dry, very firm when moist, and very sticky when wet.

For engineering purposes, more quantitative measurements of soil consistence can be made in a laboratory and expressed as a percentage of water by weight remaining in the soil when the soil displays the following characteristics:

1. *Plastic-limit* gives the moisture content when the soil crumbles as it is rolled into a "wire" between the palm of the hand and a frosted glass plate.

2. *Liquid-limit* gives the moisture content at the point when the soil flows in a curved-bottom dish after 25 impacts in a simple machine that lifts the dish a short distance and lets it drop on a hard surface. A specific tool has been designed for this measurement.

3. *Plastic-index* is the difference between the values of plastic-limit and liquid-limit.

These values are used to predict the relative ease or difficulty of working with earthen materials under differing degrees of wetness.

Effect of Humus on Physical Characteristics

Humus plays a major role in the structure of soils. Without humus, soils with a significant percentage of silt and clay become very dense and cloddy when they are tilled repeatedly (Fig. 3.11). Humus promotes the formation of soil aggregates, which are clusters of mineral grains held together by a combination of clay and humus. A well-aggregated soil is said to have good *tilth* because its looseness allows movement of roots, air, and water and resists the formation of soil crusts that can prevent the emergence of seedlings.

Water readily enters a well-aggregated soil. More water is stored in the pore systems of well-aggregated rather than poorly aggregated soils, and it is in a form that is more available to plants. Humus therefore improves water availability by promoting formation of aggregates. The attraction of water by humus is of less importance.

Well-developed structure is an indication of a high-quality soil. In all likelihood such a soil has an ample supply of humus and, if farmed, has been maintained properly. A structured subsoil promotes the penetration of water, air, and plant roots.

Poor soil management can lead to the breakdown of soil structure. This can happen by farming the soil without replacing the organic matter that is lost through decomposition. Pulverizing the soil destroys aggregates in the plow layer. Tilling the soil when it is wet is especially harmful to its structure because aggregates are easily crushed, so the soil becomes cloddy. In this form, the soil is said to be "puddled." Freezing helps to rupture large clods, but maintenance of good soil structure is the result of proper long-term farming practices.

Soil Color

In Chapter 2 we considered the differences in the appearance of the soil from one region to another. The color changes reflect, for the most part, differences in the quantity of humus and the chemical form of the iron present. It is true, how-

3.11. Soil without humus becomes cloddy (*left*), whereas humus-rich soil is granular (*right*).

ever, that the pigmentation of a given amount of humus is usually darker in grassland regions than in forested regions, particularly in warm areas.

Color of the subsoil gives a strong indication of soil hydrology and, in some cases, a high concentration of carbonates and/or sulfates.

Varying shades of red, yellow, and gray in soils are usually due to the concentration and form of iron present. Red means that the iron is oxidized and not hydrated. Yellow indicates hydration and sometimes less oxidation. Gray indicates chemical reduction caused by wetness and lack of oxygen. An exception to this is the gray E horizon just below the surface of some well-drained soils.

Gray colors in the subsoil or a combination of gray and blotches of yellow and red mottles are extremely important for interpreting the natural drainage condition of the soil. Mottles are found at a depth to which the water table rises periodically during the warm seasons. Even when the water table drops, telltale signs of soil colors are left behind, and these are used as a basis for designing septic systems, tile drainage, and the like.

The absorption of solar radiation is greater on dark surfaces than on light ones. This is certainly true for bare soils. Color differences have a comparatively minor effect on the temperature of the soil below the shallow surface layer but even this can be important for seed germination. Soil radiation has a greater impact on bare soils than on soils with plant cover because when soils become vegetated, leaves intercept the solar radiation before it reaches the soil surface.

Soil scientists use a set of standardized color charts to describe soil colors. These charts are called the *Munsell colors*. They consider three properties of color—hue, value, and chroma—in combination to come up with a large number of color chips to which soil scientists can compare the color of the soil being investigated. This system is superior to using descriptive terms alone, which may not mean the same thing to everybody.

CHAPTER 4

Soil Biology

Bacteria, fungi, worms, insects, small mammals, and many other organisms inhabit the soil. They participate in and regulate many physical and chemical processes. Soil organisms create favorable conditions for the growth of plants and also decompose plant and animal remains.

Animals in the soil make openings through it that influence the movement of water and air into and throughout the soil. Termites air-condition their mounds by channeling air through them. A bird in Australia called the thermometer fowl piles up decomposing leaves to make a compost heap into which it places its eggs to keep them warm until hatching. The bird regularly tests the temperature of the heap with its bill and moves the eggs to the most favorable part from time to time. Penguins in Antarctica make a soil for microorganisms out of guano, eggshells, feathers, and bones.

Even the most desolate landscapes on earth have primitive soils, showing the effects of water providing for life in the soil and the translocation of salts and other compounds. There is no soil without life and no higher forms of terrestrial life without soil.

In most landscapes, plant roots extend down through the soil for several feet or meters. Aboveground parts of plants in some forests extend more than 100–200 feet high (30–60 m). Shade from the vegetation lowers the soil temperature compared to that of bare soil exposed to full sunlight. Roots take up large amounts of water, and this water is conducted up the stems to the leaves from which it passes into the air as water vapor (*see* Fig. 6.12). The many tons of plant tissue per acre that die each year—including roots, leaves, fallen branches, and bark—become a part of the soil again through decomposition by soil organisms. On well-drained uplands, leaves that fall on the forest floor at the end of the growing season in humid temperate regions are nearly all decomposed by the end of the next growing season. In nearby lakes and wetlands, decomposition of plant remains is slowed because the cover of water excludes oxygen. Plant material may accumulate in wetlands as peat (which is made up of identifiable plant parts) and muck (which is a soil composed of highly rotted, dark organic matter). In upland mineral soils this dark material is called *humus* (Fig. 4.1).

Soils are classified as mineral soils and organic soils. The difference is in the amount of organic matter present. Arbitrarily, we say that about 25% organic matter by weight is the dividing point. Soils with more organic matter are called *organic soils* (peat or muck). Soils with less are called *mineral soils* because they are composed mostly of bits of minerals and rocks. A given volume of organic matter is much lighter than an equal volume of mineral soil. Thus, a soil with 5% organic matter by weight has about 10% organic matter by volume.

Humus

Upland soils have not developed in wet boggy areas so, therefore, consist largely of mineral particles. Even so, the surface soil, or plow layer, contains considerable organic matter, which is the partially decomposed residue of plants and animals that live in the soil. Figure 4.2 shows patches of this humus in pores between roots and particles of mineral soil. Humus gives soil the dark color widely associated with high fertility, although this assumption is not necessarily true for soils that have been heavily cropped or for naturally infertile soils. In most surface soils of temperate humid regions, the humus content is between 1% and 4% by weight (or twice that by volume); but this small quantity has a great influence on the physical, chemical, and biological reactions that take place in the soil. In arid regions, the surface soil typically has less than 1% humus by weight because temperatures are favorable for organic matter decomposition and vegetative growth is limited by low rainfall.

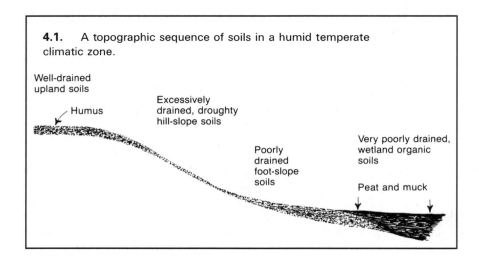

4.1.   A topographic sequence of soils in a humid temperate climatic zone.

Well-drained
upland soils

Humus

Excessively
drained, droughty
hill-slope soils

Poorly
drained
foot-slope
soils

Very poorly drained,
wetland organic
soils

Peat and muck

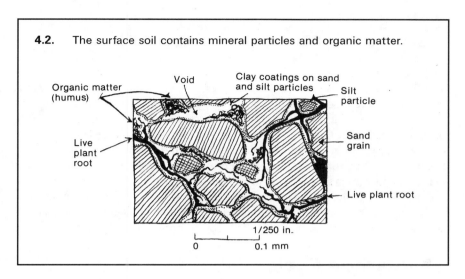

4.2. The surface soil contains mineral particles and organic matter.

Most humus is derived from roots of plants (flora) in the soil or leaf litter on the forest floor (Fig. 4.3), with only a very small fraction coming directly from soil animals (fauna). Soils formed under prairie grasslands generally have greater amounts of humus than those formed under forest vegetation because of the high density of grassland vegetation and the fibrous root systems of grasses. In the forest the vegetation at ground level is not nearly so dense, and most of the organic residue from living plants accumulates on the surface of the ground as leaves in the autumn. Much of the humus in the surface 5 or 6 inches (12 or 15 cm) of for-

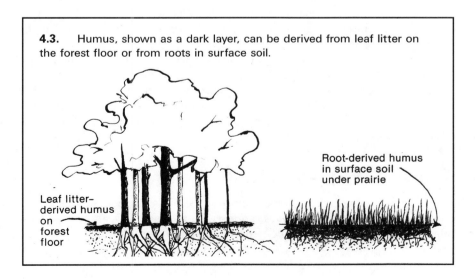

4.3. Humus, shown as a dark layer, can be derived from leaf litter on the forest floor or from roots in surface soil.

est soils results from incorporation of this plant residue into the soil by insects, worms, and other soil fauna. In agricultural use, the incorporation of plant residue and manure contributes to the formation of humus, but the decomposition of plant roots has been found to be more important.

The soil is teeming with many forms of life, each occupying a niche that is vital to the entire scheme of life. For microorganisms and small animals, the soil provides environments where conditions of feast or famine may occur side by side or follow each other rapidly. As a moist growing season is succeeded by a dry or cold season, vast numbers of organisms die. Within the soil at any one moment, small chambers full of rich humus and debris may be separated by volumes of soil that, like underground deserts, are nearly devoid of decomposable organic matter. To survive, therefore, most soil organisms must find something to eat within a few millimeters. Earthworms, on the other hand, are strong enough to make channels and move several feet in search of fallen leaves or other plant debris.

The Carbon Cycle

Life is essential to the existence of a true soil. Of the countless microorganisms that live in the soil, all but a few derive their energy from the oxidation of carbon just as humans do. Soil organic matter, most of which is humus, serves as food for these organisms. Many functions in the release of nutrients to plants are carried out by soil microorganisms, but we consider only the carbon cycle at this point.

During photosynthesis, plants take carbon dioxide (CO_2) from the atmosphere and combine it with water to produce sugar and subsequently all plant tissue. The plants die or are eaten by animals, and the residue is returned to the soil. Some of this residue decomposes on the surface, and some becomes incorporated into the soil. Ultimately, nearly all the organic material is decomposed by soil organisms and returned to the atmosphere as CO_2, where it can again be used by plants (Fig. 4.4).

Humus is continuously being decomposed and new humus is being formed to replace the old, except where human mismanagement interrupts the cycle by depriving the soil of plant or animal residues and thus impoverishing it. The soil—which supports living plants, animals, and humans—is ever ready to take into itself anything that has died. Any great tree is destined some day to fall and be incorporated into soil again.

Carbon Sequestration

One environmentally important function of soil is the sequestration of carbon through plant growth. *Sequestration* is the taking of CO_2 from the atmosphere and storing it in the organic form. The organic compounds that contain carbon exist in growing plants or as plant residues, which eventually become soil organic matter.

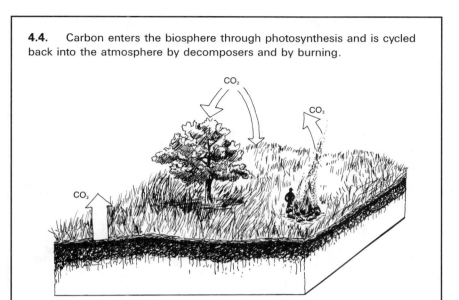

4.4. Carbon enters the biosphere through photosynthesis and is cycled back into the atmosphere by decomposers and by burning.

The organic matter content of soil has generally decreased due to the conversion of grasslands and forests to croplands, resulting in the release of CO_2 into the atmosphere. Emission of CO_2 by the burning of fossil fuels and from other sources has been the principal source of CO_2 increase in the atmosphere. The net effect of increased atmospheric CO_2 is not known, but it has been suggested that reduction of CO_2 in the atmosphere would be "environmentally friendly."

To reduce CO_2 in the atmosphere, the adoption of any practice to increase soil organic matter would be helpful. The general trend of increased crop yields over the past several decades has been beneficial. Adoption of widespread soil conservation practices has been helpful. Trees are especially beneficial because they sequester carbon in wood for long periods.

Management practices suggested to increase soil organic matter include (1) conservation tillage, (2) proper use of crop residue, such as stubble mulching, (3) application of organic materials and manures, (4) rotations to include forage or high-residue crops (such as sorghum), (5) precision agriculture, including variable rate application of fertilizer, (6) cover crops, and (7) agroforestry in which crops or forage are grown between the rows of trees. In summary, any practice is desirable if it increases yields and/or increases the amount of carbon sequestered from atmospheric CO_2.

Plant Roots and the Rhizosphere

The *rhizosphere* is the volume of soil, water, and air (with associated organisms) immediately around the root of a plant. Figure 4.5 shows that the surface of a

4.5.   The rhizosphere is the volume of the soil, water, and air immediately around the plant root.

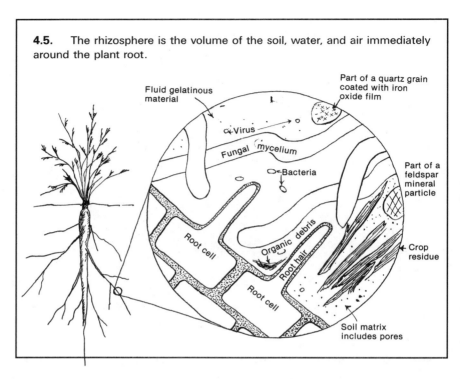

root is commonly surrounded by gelatinous material in which clay, organic debris, and microorganisms are abundant. Plant roots take up water and nutrients from the rhizosphere. The roots may release CO_2 and even oxygen. The CO_2 makes the soil solution slightly acid so that plant nutrients are more readily available for uptake. The oxygen may favor precipitation of iron to form a film in the soil near the root. Outer layers of the root may slough off, enriching the soil with organic matter.

Microorganisms

Soil microorganisms decompose and dispose of plant and animal remains. In the process these organisms form humus (Fig. 4.6), which is a more active component of soil than mineral clays. Microorganisms also perform important steps in various nutrient cycles and in solid, liquid, and gaseous phases of the soil–plant root system. Without these organic processes the cycles would lose life support. There is a biological rule stating that the smaller the organism, the greater its number and influence. Thus, the action of microorganisms in the soil is far more widespread and of greater importance than that of insects and rodents.

Some members of each group of organisms perform specialized functions in the soil. It is beyond the scope of this book to discuss all of these, but some are considered in the following paragraphs.

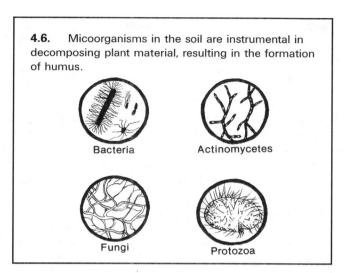

4.6. Micoorganisms in the soil are instrumental in decomposing plant material, resulting in the formation of humus.

Bacteria

Actinomycetes

Fungi

Protozoa

Living organisms are separated into the *prokaryotes*, which are not clearly either plants or animals, and the *eukaryotes*, which include the plants and animals. Bacteria and actinomycetes are in the former group; most fungi and all the protozoans are in the latter.

Estimates of the numbers of soil microorganisms in a gram of soil (about the volume of a lima bean) range from several hundred million to a few billion. Most are beneficial to agriculture, but all groups contain those that can cause crop diseases.

Types of Microorganisms

Bacteria are one-celled plants that are the most abundant forms of life in most soils. In cropland they are primarily responsible for the decay of residue but also perform a multitude of other functions. Those involved with nitrogen are discussed in more detail in the discussion of nitrogen fixation. **Actinomycetes** are mycelial bacteria that have thread-like extensions that are all part of the single cell. There may be several million actinomycetes in a gram of prairie soil. They prefer warm, dry soil, and their numbers do not diminish as rapidly with depth as those of other bacteria. The earthy or musty odor of soil comes from actinomycetes. One kind, the streptomycetes, has the antibiotic properties that we depend on so heavily in many medicines.

Algae frequently live harmoniously or even symbiotically with cyanobacteria (formerly classified as blue-green algae) to form a microbial crust on barren soils. In the United States, these crusts are particularly well developed on arid deserts of the Southwest. Cyanobacteria can fix atmospheric nitrogen (to be discussed under the nitrogen cycle), protect the soil surface from erosion, and create a favorable environment for seed germination. They also fix nitrogen in rice paddy soils and thereby fertilize the growing crops.

The **fungi**, which are most important in the soil, are multicellular organisms ranging in size from microscopic to the large mushrooms normally found only on moist, untilled soil. Fungi common in the soil are made up of a mass of fibers called a *mycelium*. One hundred thousand fungi may be found in mycelial and spore forms in a gram of soil. They do best in acid soil (pH 4.5 to 5.5), so they do not compete with bacteria, most of which flourish in nearly neutral soil. Fungi can decompose a greater variety of organic compounds than bacteria. Some catch nematodes in a kind of noose and consume them.

Besides free-living fungi in the soil, there are **mycorrhizae** that live symbiotically with roots in the surrounding soil. Threads (hyphae) of the mycelia extend into the roots of perhaps half the kinds of higher plants, which means that these plants have "double roots" of high efficiency. The hyphae feed water and nutrients into the plant roots, which in return protect and in part nourish the fungus. In this symbiotic relationship the fungus may even provide some antibiotic protection to the roots. Mycorrhizae form a sheath around the plant root and either extend hyphae into the spaces between the root cells or extend the hyphae into the cells of the root, where they are finally digested. There are two types of mycorrhizae: *endomycorrhizae* and *ectomycorrhizae*. The endo group is associated primarily with field crops such as corn, rice, and alfalfa plus a few trees such as apple and citrus. The ecto group is associated mostly with trees, a common one being pine.

Another symbiotic relationship develops between fungi and blue-green algae to form **lichens**. These primitive plants can survive on bare rock because they fix atmospheric nitrogen and can extract a few nutrients from the minerals of the rock.

Protozoa are one-celled animals. There may be thousands of them in a gram of moist, humic soil. They live inside the films of water that cover soil particles. If the films dry up, the protozoa change into a resting form in which they survive until the next rain. Protozoa include amoeboid, ciliate, and flagellate forms. They contribute to the breakdown of organic matter, and some feed on tremendous numbers of bacteria, thus helping to maintain the balance of nature.

The **myxomycetes** are slime molds, which are intermediate between protozoa and fungi. In the protozoan stage the cells are free-living. In the fungal stage they come together to form a jellylike mass that may be orange, purple, or some other bright color. The fungal stage produces reproductive spores.

The Nitrogen Cycle

Most nutrients, such as phosphorus, calcium, magnesium, and potassium, are derived from minerals, taken up by plants, and built into living tissue. The plants die and return to the soil, where they decompose and release the nutrients, which can be taken up by plants again. This is a common nutrient cycle. In the case of nitrogen, minerals are not the source, but rather it is the atmosphere, which consists of 78% nitrogen in gaseous form. In the nitrogen cycle, nitrogen goes from the air to special bacteria that exist in the soil and in roots of some plants, through several transformations, and back to any plant that absorbs it. Under certain conditions it may go back to the atmosphere instead of to plants.

The steps of the nitrogen cycle are shown in Figure 4.7.

Nitrogen Fixation. Nitrogen fixation is a process wherein nitrogen is taken from the soil atmosphere and converted into protein in living cells. Nitrogen fixation may be *symbiotic* or *nonsymbiotic*. In the case of symbiotic fixation, bacteria live in root tissue of plants to the mutual benefit (symbiosis) of themselves and the host plant. The bacteria supply themselves and the host plant with nitrogen, which is built into protein molecules. The host plants supply the bacteria with nutrients and moisture. Small knots of tissue called nodules form on the roots when these bacteria are present. Legumes such as clover, alfalfa, peas, beans, and locust are primary hosts for symbiotic nitrogen-fixing bacteria (rhizobia) (Fig. 4.8).

A vigorous alfalfa crop may fix 100–200 pounds of nitrogen per acre (110–220 kg/ha) per year, which is one reason for its inclusion in a crop rotation. Most grasses, including grain crops, are not natural hosts for nitrogen-fixing bacteria. Scientists are trying to find ways of breeding new varieties that can function in this way. If they are successful, the resultant varieties may reduce the use of commercial nitrogen fertilizer for such crops as corn, wheat, oats, and barley.

Symbiotic nitrogen fixation is also brought about by an actinomycete in the genus *Frankis* in association with several woody plants, particularly alder, Russian olive, and sweet fern. The amount of nitrogen fixed in this manner is comparable to that fixed by the leguminous crops discussed above.

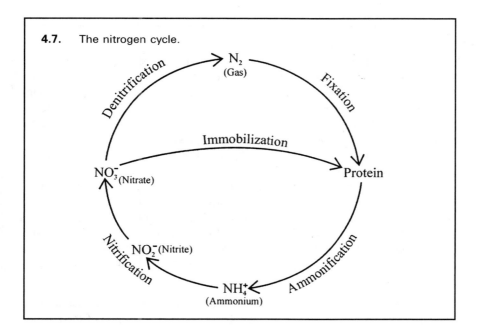

4.7. The nitrogen cycle.

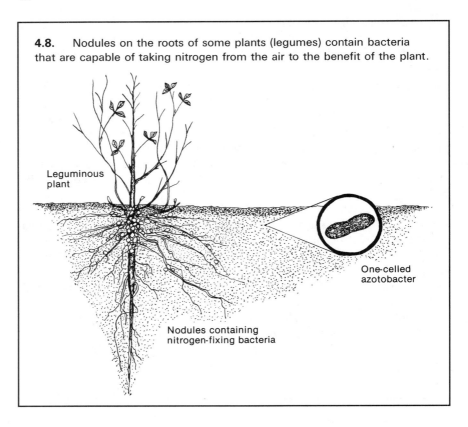

4.8.   Nodules on the roots of some plants (legumes) contain bacteria that are capable of taking nitrogen from the air to the benefit of the plant.

Leguminous plant

One-celled azotobacter

Nodules containing nitrogen-fixing bacteria

Some nitrogen is fixed nonsymbiotically. A free-living soil bacterium (Fig. 4.8) of the genus *Azotobacter* fixes nitrogen that becomes available to plants when the bacterium dies. The amount of nitrogen fixed in this manner is seldom more than 10 pounds per acre (11 kg/ha) per year but is a valuable part of the nitrogen cycle. The cyanobacteria, discussed above, also fit into this category.

In the three types of nitrogen fixation discussed above, the end product is protein, which, in this form, is not usable to any significant extent by the other plants in the ecosystem.

Some nitrogen fixation takes place in the atmosphere, particularly by lightning during storms, and the nitrogen is washed into the soil by rain. This is a different type of nitrogen fixation in that the end product is one of the oxides of nitrogen that need not progress through the nitrogen cycle to be available for plant uptake. The amount of nitrogen fixed in this way may average 10 pounds per acre (11 kg/ha) per year. Rain also carries down some nitrogen oxides from the engine exhaust of cars, trucks, and tractors.

Ammonification. Following nitrogen fixation, the next step in the nitrogen cycle is ammonification. This is the microbiological decomposition of the protein-rich root nodules or any part of the host plants of the rhizobia. Ammonia (NH_3) is a product of this decomposition and it completely ionizes to ammonium (NH_4^+) in the soil solution. Some ammonium ions may be taken up by plants in this form, but most will progress through the next two-part step, nitrification, if the soil is warm and moist. In the case of paddy rice, ammonium ions are the main source of nitrogen.

Nitrification. If the soil is warm and well supplied with moisture and oxygen, the ammonium ions are oxidized first to the nitrite (NO_2^-) form by the bacterium *Nitrosomonas.* The nitrite form rarely accumulates in the soil because further oxidation to the nitrate (NO_3^-) form is brought about directly by the bacterium *Nitrobacter.* Nitrate is the highest oxidation state for nitrogen. It is very soluble and is subject to leaching. If water percolates through the soil, nitrate moves with it and may contaminate groundwater.

Immobilization. Immobilization is the conversion of ionic inorganic nitrogen into organic nitrogen. After nitrification, nitrate ions may be taken up by the roots of higher plants or by microorganisms decomposing organic residues in the soil. This process is responsible for the nitrogen nutrition of crops and other vegetation. It is during this process that an overabundance of microorganisms may out-compete crops for the available nitrate. This may cause the crops to be nitrogen deficient if there is too much straw from a previous crop incorporated into the soil and inadequate time for its partial decomposition prior to the next crop.

Denitrification. Denitrification is the process whereby nitrate nitrogen undergoes chemical reduction and is volatilized back into the atmosphere as gaseous N_2. This process is caused by anaerobic microorganisms that flourish in saturated soil and derive their oxygen from the nitrate ions or similar oxides. To a farmer, this can be a costly loss of an expensive plant nutrient—nitrogen. This is a major reason for maintaining adequate soil drainage and proper timing of nitrogen fertilizer application.

Biological Decomposition of Rocks

Three years after the island of Krakatoa was largely blown away by a violent volcanic eruption in 1883, scientists visited it and found that the surface of the fresh bedrock was already being invaded by cyanobacteria, one of the most self-supporting form of life on earth. It can both photosynthesize and fix nitrogen. Growing along with the cyanobacteria were nitrogen- and carbon-fixing bacteria as well as fungi and lichens. Weak acids produced by these microorganisms were dissolving nutrients (phosphorus, calcium, and so on) from the rocks and building up a humic mat capable of supporting mosses and eventually higher plants. The weak acids include carbonic acid formed by solution of carbon dioxide gas in water and lactic acid produced by fungi, and the stronger acids (nitric and sulfuric) were formed by bacteria. Certain fungi and bacteria can release phosphorus from mineral particles. It is evident that microorganisms are involved in rock weathering from the start.

Macroorganisms

Most of the wild animals of the world live in the soil. In an acre of soil there may be a million nematodes, a million ants, two hundred thousand mites, and four thousand worms, to name just a few. Most wild bees nest in the soil and in the process make the soil more porous by excavating burrows and chambers. Before settlement by European immigrants, a squirrel could cross the state of Ohio without touching the ground. Many of the trees in the native forests were planted by squirrels. Obviously, animal life has greatly influenced both plants and soil.

Worms

Nematodes are eel-shaped, unsegmented, colorless worms that are abundant in the soil (Fig. 4.9). Most are too small to be seen without a microscope, but some may grow to a centimeter or more in length. Many are saprophytic, which means that they feed on dead plant residue, but some are parasitic and live on the roots of plants. They cause great economic loss to many crops, including citrus, cotton, soybeans, alfalfa, corn, and garden vegetables such as potatoes and tomatoes.

Earthworms (Fig. 4.10) the world over perform an important function in mixing organic matter with mineral matter. In a sense, they are soil factories. Among the many kinds of earthworms, the nightcrawler, *Lumbricus terrestris*, was brought to the United States from Europe by settlers. Figure 4.11 shows a tuft of dead leaves and leaf parts pulled into a burrow by an earthworm. The tuft is called an earthworm midden. In moist summers, vast quantities of leaves on the forest floor are pulled into the soil by earthworms, thereby enriching it and making it more porous. As a result of earthworm activity, the thick mats of tree leaves that accumulate on the forest floor in the autumn may be almost completely incorporated into the soil by the following year. Many other kinds of earthworms do not make middens but in their own way incorporate organic matter into the soil. In general, worms perform an important aeration and mixing function by burrowing through the soil, consuming organic matter, and bringing the residue to the surface as castings. It is estimated that worms bring 7 to 18 tons of soil per acre (16 to 40 metric tons/ha) annually to the surface in this way.

It is estimated that in a sugar maple forest 200,000 *Lumbricus* earthworms dying a natural death release into the normally acid soil a half ton of small calcite

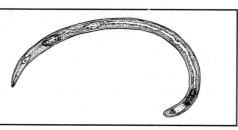

4.9.   Nematodes are usually microscopic, but they can be destructive to crops.

4.10. Earthworms are essential for mixing organic material with minerals in the soil.

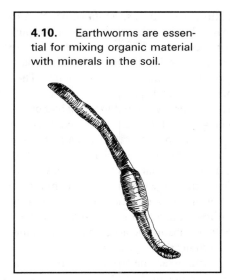

4.11. Earthworms pull parts of leaves into their burrows, thus enriching the soil.

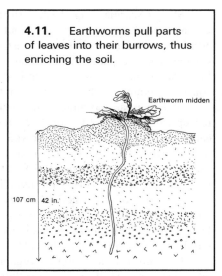

Earthworm midden

107 cm | 42 in.

nodules per acre (0.5 metric tons/ha). They further enrich the soil with the decomposition products of their soft bodies. Heavy early autumn rains may wash forest litter, worms and all, down steep wooded slopes, adding considerable fertility to downslope soils.

Arthropods

Springtails (Collembola) are primitive insects that do not go through stages of metamorphosis as flies and butterflies do (Fig. 4.12). They look like ancient fossil creatures. They can catapult themselves into space by means of a spring-like appendage on the underside of the abdomen. They are numerous in decaying leaves, and in late winter they appear on snowbanks (hence their nickname "snow fleas"), where they feed on scattered pollen. Collectors find springtails wherever there is soil, even in turf strips along freeways leading into a big city like Chicago.

4.12. Springtails and mites play an important role in the decomposition of dead leaves and stems.

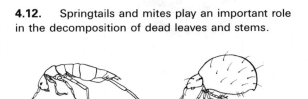

Springtail (Collembola)   Mite (Acarina)

Mites (Acarina) (Fig. 4.12) perform the same job as springtails, which is to consume dead and decomposing plant parts. Look with a hand lens at the pine needle mat under a coniferous forest for specimens. The mites in the mat are smaller than the parasitic ticks that attach themselves to people and animals. Mites are found everywhere, even in ocean depths and on high mountains. They consume organic residues and feed on nematodes and springtails.

In both urban and rural environments, ants are active in tunneling and bringing up subsurface soil to construct mounds of various sizes. Because ants can carry particles no larger than allowed by the gap between their open mandibles (mouthparts), the mounds contain no stones or gravel. Figure 4.13 shows a cross section through a mound nearly 1 foot (30 cm) high that was built by the western mound-building ant, *Formica cinera*. These insects were originally common in the grasslands of the American prairie but are now confined by cultivation to undisturbed lands such as those along railroad tracks, cemetery edges, and wetland borders. Their mounds are built largely of subsoil and are rich in organic materials that the ants bring to the colonies from nearby vegetation.

In subtropical and tropical regions, termites are particularly active in soil and plant materials. They consume large quantities of dead trees and shrubs and plant debris. Some of these insects tend to concentrate nutrients such as calcium

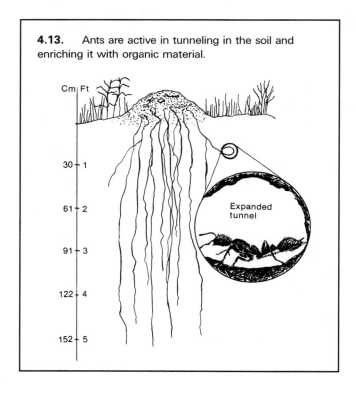

4.13.   Ants are active in tunneling in the soil and enriching it with organic material.

Expanded tunnel

in their nests, which, when abandoned, are cultivated by farmers and eventually produce patches of high-quality crops. In semiarid regions, underground termite nests may act as sinks (collectors) for irrigation waters and thereby become a nuisance. Hard aboveground mounds in cultivated fields are undesirable obstructions. The long-term soil-mixing effects of termites are beneficial, but the immediate effects may be troublesome. Some mounds may be higher than those shown in Figure 4.14.

Vertebrates

Moles (Talpidae) plow soil by burrowing just below the surface to where they can find earthworms, grubs, and plant roots to eat. This activity occurs both in sod and in forest topsoil. It leaves the soil loosened and contributes to the high porosity of noncultivated soils.

Mice (Cricetidae) and shrews (Soricidae) are numerous enough to make an impact on soils by their burrowing activities. When snow melts in the spring,

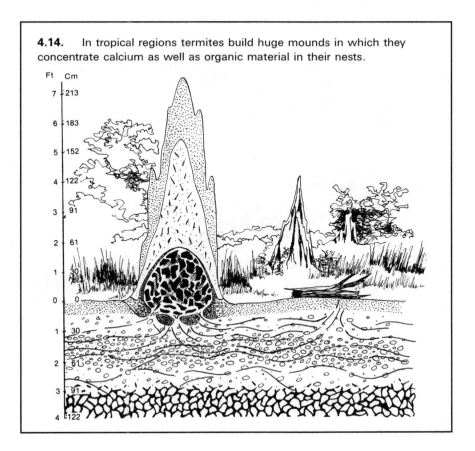

4.14. In tropical regions termites build huge mounds in which they concentrate calcium as well as organic material in their nests.

networks of rodent runways are plainly visible. Ground squirrels (*Spermophilus*), marmots (*Marmota*), prairie dogs (*Cynomys*), and other mammals make elaborate burrows, constructed to not fill with water readily during rainy periods and to be aerated by convection and updraft air currents. These rodents bring tons of subsoil material to the surface. Because these animals prefer dry sites, the materials they excavate are commonly sandy and gravelly. The soil profile is locally quite churned up and enriched with vegetable debris and rodent excreta in the process (Fig. 4.15).

Pesticide Use—A Dilemma

Before the dawn of agriculture, all organisms were in balance and none were able to build up in numbers beyond that of natural populations. This is not to say

4.15.   The burrowing activities of animals contribute to the porosity and enrichment of soils.

that primitive humans were not bothered by insects and the like, but theirs were natural populations. When humans began to manipulate plants and animals to increase their food supply, the balance was altered so that certain organisms became detrimental to the aims of agriculture. Various forms of control have been used, but in recent decades the emphasis has been upon organic compounds that are intended for the selective control of specific target organisms (Fig. 4.16). Some pesticides kill certain kinds of pests such as fungi, nematodes, insects, or rodents; some regulate plant growth by speeding it up or retarding it; some defoliate or desiccate plants; some attract insects to deadly traps or sterilize them; and some repel pests through protective coatings such as are found on some seeds.

Some of the problems with insecticides have been instrumental in alerting the public to the hazards of pest control. Those that bacteria quickly break down into harmless components are deactivated and, when properly handled, cause little or no injury to the environment. Those that are unlike any natural molecules cannot be attacked by bacteria; thus, they can build up in the environment. DDT is an example of an insecticide with low biodegradability that was once used heavily worldwide but is now banned from use in the United States and many other countries. There are other examples where a component formed by microbial breakdown of the original insecticide is potentially very hazardous to human health. One of the aldicarbs derived from a combination insecticide-nematocide used on potato fields is in this category. Aldicarb (or its derivatives) has been found in the groundwater beneath sandy soils in Wisconsin, New York, and Florida. If the soil is not rapidly permeable, the breakdown of aldicarb into harmless components seems to be complete.

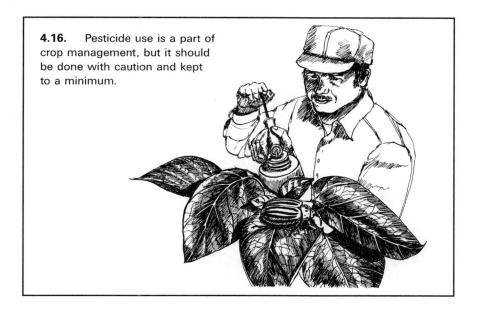

4.16. Pesticide use is a part of crop management, but it should be done with caution and kept to a minimum.

The massive use of pesticides over large areas has, in some cases, been self-defeating. Sometimes natural enemies of a pest have been eliminated, and pesticide-resistant varieties of the pests have evolved. Well-planned harmonizing of chemical and natural control methods (integrated pest management) is a wiser approach. Limited, strategic use of pesticides may be combined with ecological pest control. The latter includes the encouragement of growth of populations of natural enemies of pests, release of many sterile individuals of a species, and rotation of crops in a way so as to interrupt population expansion. The tobacco horn-worm moth, for example, has been controlled by light use of pesticides together with a vigorous encouragement of parasitic wasps (biological control) and some handpicking (scouting) of larvae. Integrated pest management has been well received because it is an economically sound approach as well as being good for the environment. The recent development of hybrids that are resistant to specific insects and infections also offer an opportunity to reduce the application of pesticides. Some examples are corn that is resistant to corn borers, potatoes that are resistant to potato beetles, and alfalfa that is resistant to leaf hoppers. The application of insecticides to control these and other insects has been extremely expensive and controversial.

CHAPTER **5**

Soil Colloids
and Chemistry

The colloidal system of the soil is made up of very fine clay and humus particles that have negative and positive electrically charged sites. Colloids are too small to be seen with a light microscope, but clear images of them can be made with electron microscopes. The upper limit of their diameter is commonly given as 0.0001 mm although particles somewhat larger may react similarly but to a lesser extent. For comparison, it would take 254,000 of these particles, side by side, to extend 1 inch (2.54 cm). Thus, the colloidal system is made up of the finest clay particles and highly decomposed humus. Colloids are the most chemically active fraction of the soil and are intimately associated with many reactions involved in plant nutrition.

Since colloidal clay and humus particles have negatively and positively charged sites, nutrient ions that are essential for plant growth are attracted to the colloidal surfaces of opposite charge. The positively charged ions are *cations* and those with negative charge are *anions*. They are weakly held as a reserve supply for plants to draw upon. Without the attraction between ions and colloids, the leaching of ions deeper into the soil and beyond the reach of roots would be much greater in humid regions. The nature of the colloidal system is not only dependent on the colloids themselves but also on the properties of the ions attracted to them. These attracted ions may be exchanged, partly in accordance with the dominance of specific ions in the soil solution. This process is called *ion exchange*. In all soils, except some in tropical regions, the negatively charged sites on colloidal surfaces are much more numerous than positives sites, so the usual process is cation exchange.

To understand how colloids influence soil chemistry, it is necessary to know something about their composition. Clay mineral colloids and humus colloids will be discussed separately.

Silicate Clay Mineralogy

Mineral particles such as common feldspar grains from granite are made up mostly of three elements: silicon, oxygen, and aluminum. Therefore, they are called *aluminosilicates*. Small feldspar particles slowly change to clay minerals by weathering. These are also aluminosilicates, but they are different from feldspars in two principal ways: the clay minerals have some water molecules in their structure so they are called hydrated aluminosilicates, and they have a platy or layered structure.

Just as a plant leaf is made up of distinct layers of cells, the very small, flat clay crystals are made up of definite layers of ions. Most silicate clay particles are sandwich-like, with two silica layers (silicon plus oxygen) between which is an alumina layer (aluminum plus oxygen). They are called 2:1 clays because of this arrangement. *Smectite* and *hydrous mica* are clays of this type.

Clay that is of the 1:1 type is like an open-faced sandwich. There is a single silica layer adjacent to a single alumina layer (aluminum plus hydroxyl). *Kaolinite* is a common 1:1 clay. Plates of halloysite, a variety of kaolinite, tend to curl (Fig. 5.1).

These 2:1 and 1:1 types of clays are called *layer lattice silicate clays*. The ions in each layer are arranged in lattice-like geometric patterns (Fig. 5.2). The 2:1 lattice clays have variations within the geometric pattern of ions that give rise to a negative charge on the surface. Most 2:1 clays are also *expanding lattice clays* so they absorb water between, but not within, the sets of 2:1 lattices.

As an analogy, clay particles resemble a stack of sandwiches and the expansion takes place between the sandwiches. Expanding lattice clays have a tremendous surface area because the internal surfaces are available to react with the soil solution. Clays with a 1:1 lattice do not expand because hydrogen bonding between the sets of lattices holds them together.

Another kind of clay is the oxide clay that has little or no regularity in its structure. In this respect, oxide clays are gel-like.

5.1.   Clay particles are extremely small, and in some types the layers tend to curl.

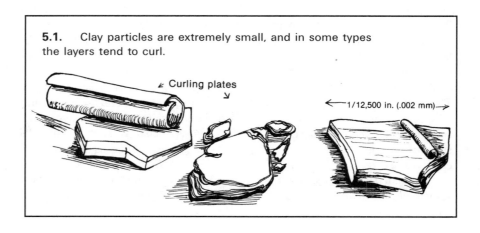

Curling plates

←—1/12,500 in. (.002 mm)—→

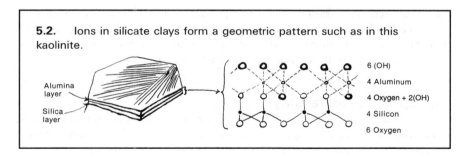

5.2. Ions in silicate clays form a geometric pattern such as in this kaolinite.

Alumina layer

Silica layer

6 (OH)
4 Aluminum
4 Oxygen + 2(OH)
4 Silicon
6 Oxygen

Source of Negative Charge on Silicate Clay Mineral

The negative charge on silicate clays in soils comes from two sources. A typical silicate clay of the 2:1 type illustrates this principle. First, the silica layer develops a negative charge from the oxygen ions along the edge of the crystal. Only one of the oxygen's two negative charges is combined with a silicon ion, so at the plane where the crystal ends, there are oxygen ions with one negative charge unsatisfied. Figure 5.3 depicts this charge distribution in two dimensions, with an unsatisfied charge at each end of the lattice. The oxygen ions are not shown in this schematic diagram, but their location is similar to the ionic arrangement in the silica layer shown in Figure 5.2. This source of negative charge is called *edge charge*, and although it is low, it is the main charge on kaolinite clay. The edge charge in allophane is somewhat higher. Edge charge is a *pH-dependent* charge because, under acidic conditions, H^+ can nullify the negative charge. Therefore, this charge is not permanent.

The second source of a negative charge arises when one ion is substituted for another during the formation of the silicate clay crystal, without any change in its form. This is called *isomorphous* (equal form) *substitution*, and it can oc-

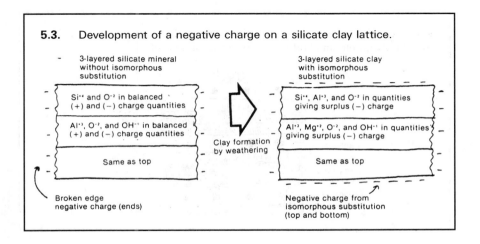

5.3. Development of a negative charge on a silicate clay lattice.

3-layered silicate mineral without isomorphous substitution

3-layered silicate clay with isomorphous substitution

Si^{+4} and O^{-2} in balanced (+) and (−) charge quantities

Al^{+3}, O^{-2}, and OH^{-1} in balanced (+) and (−) charge quantities

Same as top

Si^{+4}, Al^{+3}, and O^{-2} in quantities giving surplus (−) charge

Al^{+3}, Mg^{+2}, O^{-2}, and OH^{-1} in quantities giving surplus (−) charge

Same as top

Clay formation by weathering

Broken edge negative charge (ends)

Negative charge from isomorphous substitution (top and bottom)

cur in different ways. In some clays an aluminum ion (Al^{+++}) substitutes for a silicon ion (Si^{++++}) in the outer (silica) layers, whereas in other clays a magnesium ion (Mg^{++}) may substitute for Al^{+++} in the alumina layer. Either way, one negative charge results in the crystal and the charge is permanent.

Groups of Silicate Clays

Several groups of layer silicate clay minerals have been identified and within each group there are many specific clay minerals. In this book, only three of these groups are discussed to illustrate the nature and importance of clay.

Smectite Group. *Montmorillonite* is a common member of the smectite group. It is a 2:1-type clay with a high capacity to hold plant nutrients and to swell and shrink on wetting and drying (Fig. 5.4). Variations within this group are due mainly to the amount of substitution of magnesium and ferrous iron for aluminum in the alumina layer. Soils that have high amounts of montmorillonite clay can be very troublesome, particularly when wet. They are expanding lattice clays wherein their strong affinity for water causes the clay particles to spread apart and readily slip past one another. This results in what is called low ***bearing strength***, which means that foundations of buildings and roads built on these clays are likely to fail (slip) and cause cracking in the superstructure, particularly on sloping ground. When montmorillonitic soils dry, cracks of nearly 2 inches (5 cm) or more may open. Debris may fall into these cracks and cause the soil to buckle when it is wetted. Montmorillonitic soils become very sticky and difficult to till when wet and very hard when dry. As a result, farmers can work them only at just the right moisture content. Unimproved roads on montmorillonitic soils become impassable in rainy seasons.

The inner (alumina) layer of the montmorillonite clay lattice is made up of aluminum, hydrogen, and oxygen ions. All the negative and positive charges bal-

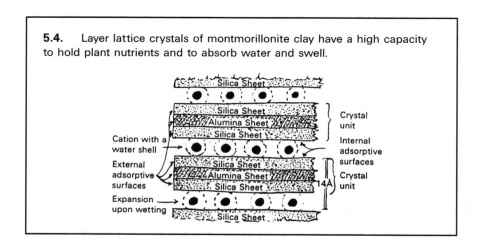

5.4. Layer lattice crystals of montmorillonite clay have a high capacity to hold plant nutrients and to absorb water and swell.

ance and neutralize each other within this layer only if the three named ions are present. In montmorillonite clay, about one-fourth of the aluminum ions (Al^{+++}) have been replaced by ions of magnesium (Mg^{++}) or iron (Fe^{++}); ions with two positive charges have been substituted for ions with three positive charges. This produces a deficiency in positive charges, which results in an excess of negative charges at the surface of the crystal lattice. These are permanent negative charges that developed when the crystals were formed.

Smectite clays tend to be associated with the subhumid to arid climatic regions that have produced grasslands in the United States. When found in the more humid regions, they are generally in soil formed from shale or in the residue from basic rocks.

Hydrous Mica Group. Hydrous mica (Fig. 5.5) has a rather slight structural difference from the primary mineral (mica) that is found in granite. Hydrous mica is probably derived by weathering of mica. It is associated with regions where weathering has not been severe and where the soil is neither very acid nor very basic. A member of this group is called *illite* after a location in Illinois where it was first identified. Hydrous mica is like montmorillonite in that it has a 2:1 lattice structure, but the lattice layers are held together by a mutual bond with potassium ions between them. This bonding minimizes the swelling and shrinking and results in good bearing strength for this clay and in reduced stickiness when wet. Illite has a lower capacity to hold plant nutrients than montmorillonite.

The presence of hydrous mica in a soil does not make the soil unstable in the way that montmorillonite does. A predominance of hydrous mica clay in a soil indicates a lack of severe weathering. Such clays are likely to be found in the cool climatic zones where precipitation is high enough to remove soluble salts from the soil.

When the interlayer potassium is completely removed by weathering, an expanding lattice 2:1 clay called *vermiculite* is formed. It does not shrink and swell

5.5. Layer lattice crystals of hydrous mica clays have a lower capacity to hold plant nutrients and absorb water.

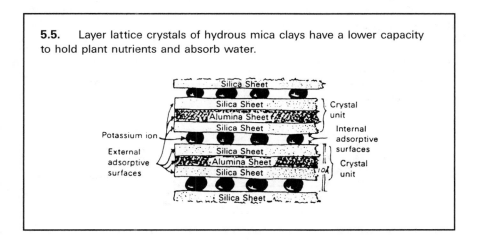

as much as montmorillonite does. In vermiculite the negative charge is derived from the isomorphous substitution of Al^{+++} for Si^{++++} in the outer layer. As a result, vermiculite has a higher negative charge than does montmorillonite.

Kaolinite Group. The lattice of kaolinite clays is a 1:1 type made up of one silica and one alumina layer (Fig. 5.6). It can be seen that kaolinite has the least silica of any of the silicate clays. This is the result of the intense weathering that is characteristic of warm regions of the world. One important property of kaolinite is the fixed spacing between the lattice layers. This is due to the attraction of hydrogen of the hydroxyl ions in an alumina layer for the oxygen in the adjacent silica layer. The bond between these lattice layers is of great importance because it renders kaolinite less sticky and gives the soil a greater bearing strength than with other types of silicate clays. Kaolinite has a very low capacity to hold plant nutrients and it absorbs less water than 2:1 clays.

Kaolinite, a favorite clay among potters, is most abundant in tropical and subtropical regions. Nearly pure deposits of kaolinite are valuable as sources for industrial materials. Large amounts are mined for use in the manufacture of bathroom fixtures.

Identification of Layer Silicate Clay Minerals

Although different clay minerals have contrasting properties, it is important to identify them so that a soil's capabilities and limitations can be accurately predicted. The most common laboratory instrument used to identify silicate clays is the X-ray machine (Fig. 5.7). X-ray technology was first applied to soils by Bragg in about 1912. It was found that rays of a very specific wavelength are reflected back to a detector when they strike parallel clay lattice surfaces as the specimen of clay is rotated in an X-ray beam. The angle of rotation needed to attain reflection depends on the spacing between the lattice layers. This property is specific for the various kinds of clay when they have undergone certain pretreatments.

5.6. Layer lattice crystals of kaolinite clay have a very low capacity to hold plant nutrients and to absorb water.

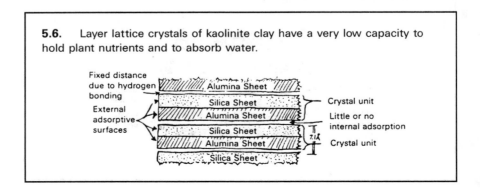

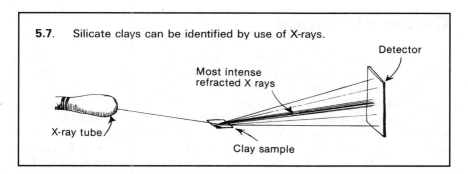

5.7. Silicate clays can be identified by use of X-rays.

Detector

Most intense
refracted X rays

X-ray tube

Clay sample

Noncrystalline (Amorphous) Silicate Clays

When volcanic ash weathers in a relatively short time, some nearly amorphous silicate clays form. Two of the clay minerals are ***allophane***, which is spherical, and ***imogolite***, which is thread-like. Except for their small size, they share few of the properties of the layer silicate clays. Their presence is germane to the classification of soils of volcanic origin.

Oxide Clays

To this point we have considered only silicate clays, but oxide and hydrated oxide clay minerals also are present in soils (Fig. 5.8). Normally, these are oxides of iron and aluminum, are amorphous, and are found most abundantly in soils formed from parent materials rich in iron and aluminum in tropical and subtropical regions where weathering has removed much of the silica from the clay fraction. Oxide clays have little or no crystallinity and very low capacity to hold plant nutrients. If iron oxides are not very hydrated, they give the soil a deep red color. Ferrihydrite is one of the iron oxide clays that is very hydrated; it is important in the classification of some volcanic region soils.

Where aluminum oxide is concentrated enough in the subsoil and below, it is called ***bauxite***, an ore of aluminum. Bauxite is mined and loaded onto ships to be sent to areas where there is cheap electrical power for processing the ore to yield pure aluminum metal.

5.8. A particle of oxide clay has little or no crystallinity and a very low capacity to hold plant nutrients.

Cation Exchange

Although most soil colloids have a net negative charge, no electrical charge in the soil goes unbalanced for very long. Electrical neutrality is maintained.

A soil colloidal system has a double layer of charges. The inner layer is the negative charge of the colloidal particle discussed above. The outer layer is formed by cations in the soil solution, which are attracted to the colloidal surfaces in proportion to the negative charges available. This means that a divalent cation such as calcium (Ca^{++}) or magnesium (Mg^{++}) can neutralize two negative charges of the colloidal particle, whereas monovalent ions such as potassium (K^+), sodium (Na^+), or hydrogen (H^+) can neutralize one negative charge each (Fig. 5.9). In acid soils, aluminum ions (Al^{+++}), which may be combined with one or two hydroxyls (OH^-), can be attracted to colloidal surfaces. There may be many other cations attracted to the colloids in small amounts. Some of these are trace elements that are of great significance to growing plants.

Cations in the outer layer are sometimes called "swarm ions" because they resemble a swarm of bees around a hive, with the greatest concentration of bees close to the hive (Fig. 5.10). In a soil colloidal system, these cations become hydrated so their effective radius includes the water molecules. The ions with two or more positive charges (such as calcium) and small effective diameters are attracted close to the colloid, whereas monovalent cations (such as sodium) with one positive charge and a large effective diameter tend to migrate farther from the

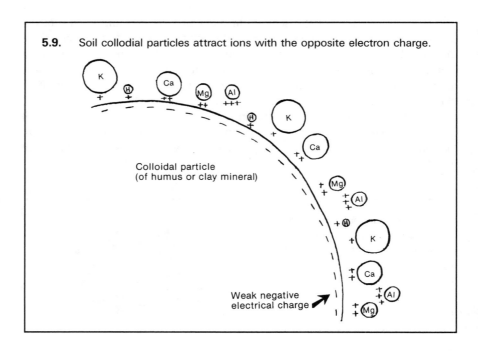

5.9. Soil collodial particles attract ions with the opposite electron charge.

5.10. A "swarm" of positively charged ions around a negatively charged soil particle resembles bees around a hive.

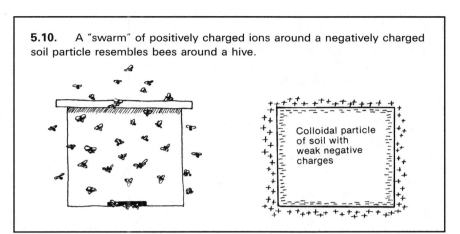

surface of the colloid. All the attracted cations are in constant motion, but attraction holds them tightly enough so they are not readily lost to water that is moving through the soil. They are *adsorbed* ions because they are held to the surfaces of the colloids. This action is very important to plant life because it keeps many nutrients within the root zone of the crops.

Cations (for example, Ca^{++}) in a mineral fragment are released by weathering into the soil solution where they are attracted to particles of clay, around which they "swarm." By exchange with hydrogen ions coming from around roots, the nutrient ions finally reach the roots. There is an area called the *oscillation zone* in which ions are moving around roots and clay particles. This is the place of exchange, where one cation is replaced by another with an equivalent amount of charge (Fig. 5.11). For example, one divalent ion (such as Ca^{++}) may replace another divalent ion (such as Mg^{++}) or it may replace two monovalent ions (K^+ and K^+). When a plant takes cations from the soil solution (Fig. 5.12), it releases hydrogen ions (H^+) in exchange. For example, when one calcium ion is taken into the plant, two hydrogen ions are given off into the soil solution. Thus, electrical neutrality is maintained.

Cation Exchange Capacity

To quantify the negative charges on the soil colloids and therefore also the amount of cations attracted to those charges, it is essential to express the amount in standard units. The units are *centimoles of charge per kilogram of soil material* ($cmol_c/kg$). The "c" subscript before the slash indicates "charge." The quantities determined are designated as the ***cation exchange capacity (CEC)***. Typically this measurement is determined on soil samples, but it may be made on other earthy deposits such as lake bottom sediments.

There are many variations in the laboratory determination of CEC, but the basic principles behind the methods are similar: (1) A known weight of soil is

5.11. A calcium ion (Ca^{++}) (*left*) migrates in solution toward a negatively charged soil particle to which two potassium ions (K^+) have been previously attracted. The Ca^{++} ion (right) changes places with the two K^+ ions, which move on into the soil solution. An instance of cation exchange has occurred.

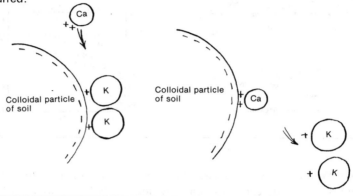

5.12. Cations move from a mineral, into solution, to the colloid surface, and on into the rootlet by ion exchange.

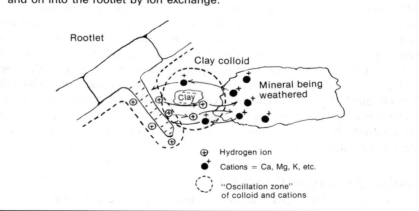

placed in a beaker and reacted with a solution containing only one type of cation, such as ammonium (NH_4^+). (2) When it has been established that all the negative sites on the colloids are satisfied with ammonium ions, the ammonium ions are replaced with another ion, and the ammonium ions replaced are measured. (3) The $cmol_c/kg$ of NH_4 determined represents the CEC of the soil sample.

Frequently the kinds of cations held on the colloidal system need to be determined. This can be done in a similar manner wherein the exchangeable cations in the soil sample are replaced with another cation and those removed are analyzed individually. The kinds of cations found on the colloidal system of most soils are quite predictable. They are calcium (Ca^{++}), magnesium (Mg^{++}), potassium (K^+), sodium (Na^+), and hydrogen (H^+). In some soils, aluminum (Al^{+++}) may also be very significant. When nitrogen is added to the soil in the form of ammonia (NH_3) or the ammonium ion, (NH_4^+), the adsorption of (NH_4^+) becomes important. The range in the CEC for pure samples of the clays discussed in this chapter are shown in Table 5.1.

Table 5.1. The range in cation exchange capacity of some common clay materials.

Type of Clay	CEC in $cmol_c$/kg
Kaolinite	3–15
Illite	10–40
Montmorillonite	80–100
Vermiculite	100–150

For most agricultural soils, the CEC ranges between 3 to 20 $cmol_c$/kg. The very sandy soils are at the low end of the scale and very clayey soils or organic soils may have a CEC much higher than 20 $cmol_c$/kg. The CEC of the soil gives a strong indication of the ability of a soil to retain and release nutrients, but it does not replace a soil test for plant nutrients to be discussed later.

Humus as a Colloidal Substance

Humus is also part of the soil's colloidal system, and it releases valuable plant nutrients as it decomposes. Like clay, these microscopic particles carry negative charges to which cations are attracted. The negative charge of colloidal humus particles can develop in several ways. One is from the migration of hydrogen ions (H^+) away from certain organic chemical groups. An example is what happens to carboxyl groups—made up of carbon (C), oxygen (O), and hydroxyl (OH^-)—along the sides of the humus colloids (Fig. 5.13). The CEC of humus is twice or more times as high as that of silicate clays, so its value to the soil for crop production is enormous. The abundance of hydrogen ions in soils with a low pH restricts the migration of H^+ from the surface of humus colloids and reduces their CEC. For this reason humus has a pH-dependent charge rather than a permanent charge.

Anion Exchange

Up to this point, we have considered only those colloids that have a net negative charge because this is usually the dominant condition. However, the word *net* in net negative charge implies that positively charged sites also may be present, though in lesser numbers, and this is indeed the case. In fact, certain acid tropical soils contain colloids with a net positive charge, and these colloids attract and exchange soluble anions just as negatively charged col-

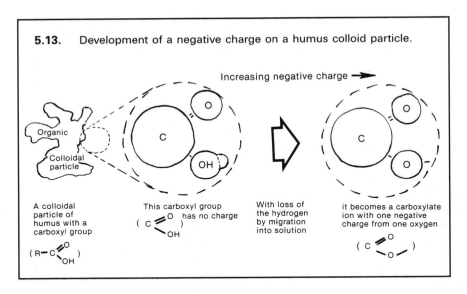

5.13.   Development of a negative charge on a humus colloid particle.

Increasing negative charge ➤

A colloidal particle of humus with a carboxyl group

$(R-C\overset{O}{\underset{OH}{\diagdown}})$

This carboxyl group $(C\overset{O}{\underset{OH}{\diagdown}})$ has no charge

With loss of the hydrogen by migration into solution

it becomes a carboxylate ion with one negative charge from one oxygen

$(C\overset{O}{\underset{O^-}{\diagdown}})$

loids attract and exchange soluble cations. Thus, these soils exhibit anion exchange instead of cation exchange. Soluble anions such as nitrates (NO_3^-), chloride (Cl^-) and sulfate (SO_4^{--}) are held and exchanged on the positively charged surfaces, whereas the cations are repelled and remain in solution. Phosphate ($H_2PO_4^-$) also is attracted to these surfaces but is held much more tightly on surfaces of iron, aluminum, and calcium-bearing minerals by a specific adsorption mechanism that operates in either positively or negatively charged soils.

Surface charge becomes more positive (or less negative) as the soil acidity increases. Soils that show positive surface charges, therefore, are characteristically acid, and their colloid component is high in kaolinite, iron and aluminum oxides, and hydroxides but low in humus and expanding-layer silicate (smectite) clay. They are usually undesirable for crop production, not so much because of their charge, but because they often contain enough active aluminum and/or manganese to be toxic to plants. Many of these soils also have a very high capacity for fixing applied phosphorus in a form that is not very available to plants. Soils high in humus and smectite clay never go positive, but some tropical soils low in these components can be positive at pH 6 and below.

From an economic and environmental standpoint, positive soils might be better than negative soils in that expensive nitrate is not as readily leached and nitrate contamination of groundwater would be reduced. However, the usual problems with aluminum and manganese toxicity and phosphorus fixation in these soils make them generally undesirable. Also, these soils are so highly weathered that very few nutrients are released by further weathering.

Soil Reaction (pH)

Soil reaction refers to the concentration of hydrogen ions (H^+) and hydroxyl ions (OH^-) in the soil solution, which are expressed in moles per liter. The term pH is a measure of the concentration and activity of hydrogen ions (H^+) in a system. It is defined as the negative log of the hydrogen ion concentration, or $-\log [H^+]$. The pH scale is the logarithm to the base 10 of the reciprocal of the hydrogen ion concentration. Thus as the pH of a solution goes from 7 to 6, for example, the hydrogen ion concentration increases 10 times and the hydroxyl ions decrease by a comparable amount (Fig. 5.14).

The pH scale extends from 1 to 14 with pH 7 being precisely neutral. This means that at pH 7 the concentration of hydrogen and hydroxyl ions is equal. As hydrogen ions increase in concentration and hydroxyl ions decrease, the pH drops below 7 and vice versa. The pH ranges that might be encountered under natural soil conditions are illustrated in Figure 5.15. It is rare to find soils close to either of the extreme ends of the pH scale unless they have been contaminated by human activity.

Soil pH is the soil chemical property most commonly measured by farmers and urban homeowners alike. The growth of crops is influenced by soil pH. Most field crops, such as corn, small grains, cotton, and pasture grasses will grow sat-

5.14.   Hydrogen ion concentration is expressed as pH.

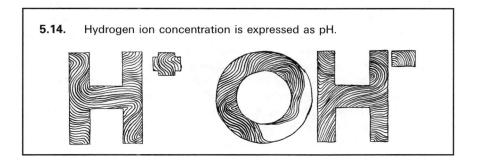

5.15.   Soil reaction is usually less than two pH units on either side of neutral.

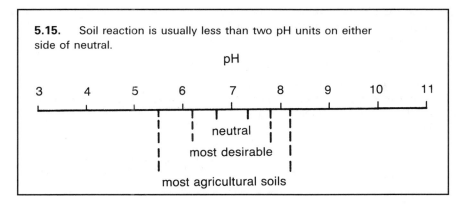

isfactorily over a relatively wide pH range—from 5.5 to 8.3—but the preferred range for best production is from pH 6.5 to 7.8. Cranberries and blueberries, however, grow best in acid soils with pH 4.0 to 5.0, while alfalfa and sweet clover require a pH of 6.5 or above for best growth.

Plants are usually not directly affected by soils that range in pH quite far either way from neutral; however, the indirect effect may be drastic. It is somewhat like having a fever that does not harm the body unless it goes to an extreme, but it certainly lets us know that we have some sort of an infection that is at least temporarily harmful. Likewise, soils that have a pH very far from neutral are likely to produce poor crops for one or more of several possible reasons such as (1) a lack of one or more plant nutrients, (2) presence of plant nutrients in forms unavailable to plants, (3) diminished activity of beneficial soil microbes, and (4) abundance of ions toxic to plants. How to correct an unfavorable soil reaction is discussed in Chapter 9.

Base Saturation

Base saturation refers to the percentage of base-forming ions that occupy the colloidal surfaces compared to the percentage of acidic ions. Most cations in the soil are associated with the cation exchange complex of colloidal clay and humus discussed earlier. Under acidic conditions comparatively few basic ions (calcium, magnesium, potassium) are present on the colloidal system, but there are many hydrogen ions. Aluminum ions on the colloidal system also give an acid reaction. In a laboratory, when the naturally occurring cations in a soil sample are replaced with another kind, the original cations can be collected and identified. The amount of basic exchangeable cations found may then be compared to the total amount of exchangeable cations. This is the means for calculating the ***percent base saturation***. For example, on the colloidal clay particle in Figure 5.16 there are 25 cations, 15 of which are basic. Therefore, the percent base saturation is $(15/25) \times 100 = 60\%$. A high percentage base saturation is usually desirable for crops.

Keep in mind that numbers given above and in Figure 5.16 are for illustration only. The actual numbers are beyond comprehension. One mole of charge corresponds to Avogadro's number (6.02×10^{23}) so, for example, 1 cmol$_c$/kg of H$^+$ would be 1/100 of Avogadro's number (6.02×10^{21}). Figure 5.17 gives a comparison of this quantity to a spoonful of soil.

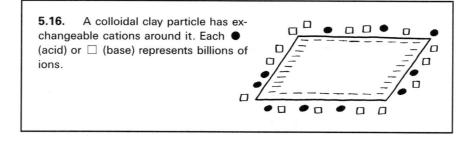

5.16.   A colloidal clay particle has exchangeable cations around it. Each ● (acid) or □ (base) represents billions of ions.

5.17. A spoonful of soil weighing 10 g (dry) contains about 1.2 quintillion (1.2 × 10²¹) exchange sites to which plant nutrients (Ca, K, etc.) can be held available for plant roots.

Buffering Capacity

In chemistry, buffering refers to a resistance to change in pH; consequently, a soil that is well buffered is one whose pH is not easily altered significantly. The cation exchange capacity of soils gives them most of their buffering capacity, as the vast majority of the basic and acidic ions are held to the surface of the clay and humus colloids and are exchangeable. Thus, when basic ions are dissolved in the solution of a colloid-rich soil, there is very little increase in pH because hydrogen ions are released from the colloidal surfaces to neutralize the base that was added. Well-buffered soils need more lime to raise their pH than those that lack an abundance of colloids. The situation is analogous to the strength of an army—it is not so much related to the number of troops it has on the front line as it is to the forces it holds in reserve.

Reasons for Basic or Acidic Soil

In arid and semiarid regions of the world, most soils are basic (alkaline) or nearly neutral for two reasons: the ions derived from weathering of minerals are predominantly base-forming ions, and there is not enough precipitation to leach them from the soil.

In humid regions of the world, leaching by precipitation causes the bases to be translocated deeper into the soil, and ultimately they return to the sea (Fig. 5.18). The

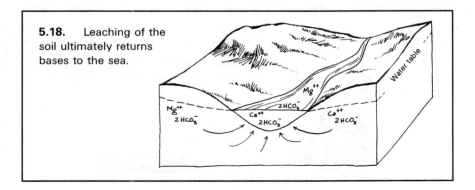

5.18. Leaching of the soil ultimately returns bases to the sea.

effect of this process over the long span of geologic time is evident in deposits of limestone and other basic sedimentary rocks laid down on the sea bottom. The limestone deposits, which may be several hundred feet thick, have resulted from the concentration of calcium and magnesium carbonates by living organisms or by chemical precipitation in ancient seas. Other basic ions have been concentrated as salt beds when seas dried up. In many places, basic sedimentary deposits are covered with thick sandstone formations that may be somewhat acid or neutral in reaction.

Some agricultural practices such as harvesting of crops and addition of organic matter or nitrogen fertilizer promote the acidification of soils. These effects, however, are not serious if steps are taken to counteract them by the addition of lime.

In an obsolete system for classifying the soils in the United States, the first separation was made between what were called the *pedocals* and the *pedalfers*. Geographically a line was drawn north and south approximately through the center of the country. To the west were the pedocals, which were basic and had an accumulation of calcium carbonate in their subsoil; to the east were the pedalfers, which were acidic and had a subsoil accumulation of aluminosilcate clay and/or iron.

Soil Aggregation

The colloidal system of the soil not only is the center of chemical reactions but also has much to do with the physical structure of the soil. Most of the common soil cations (particularly Ca^{++} and Mg^{++}) attracted to colloids cause them to cluster into what is called a flocculated condition. The opposite of *flocculation* is dispersion, which means that the soil particles do not tend to cluster together when wet. Sodium ions (Na^+) disperse soil very effectively and cause the soil to flow together so that it becomes almost impermeable to water (Fig. 5.19). The reason is that the adsorbed Na^+ migrates far enough away from the colloids to leave the negative charges of the colloids unsaturated. Like charges repel, so the negative ($^-$) particles disperse. The soil loses stability as a result. Entrance of sodium-rich water into earthen dams may cause them to fail.

The dispersed soil condition caused by sodium is very adverse to crop production. It is most commonly associated with slight depressions in otherwise level

5.19.   Soil is well aggregated by action of colloids rich in calcium ions (*left*). Soil runs together in a dense mass by action of colloids containing abundant sodium ions (*right*).

grasslands of semiarid regions as well as with cropland irrigated with water high in sodium. Sodium also brings about a strong alkali condition that can dissolve humus, producing a dark crust on the soil surface when the water evaporates. Farmers call these "black alkali spots," and they are relatively sterile (Fig. 5.20).

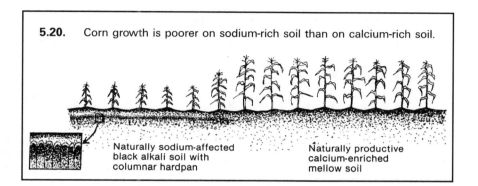

5.20. Corn growth is poorer on sodium-rich soil than on calcium-rich soil.

Naturally sodium-affected black alkali soil with columnar hardpan

Naturally productive calcium-enriched mellow soil

CHAPTER **6**

Soil Water

Many ancient civilizations left evidence that they understood how vital water was to the survival of their cultures. Nomadic tribes followed the seasonal rainfall patterns that affected the growth of forages for their grazing animals and of edible plants for their own consumption. Some of the earliest public works projects involved drainage and irrigation of lands to enhance crop production. The eventual collapse of some of these ancient civilizations has been attributed to poor management of water resources. Human reliance on a sufficient and timely supply of water for food and fiber production is no less critical today.

Water in soil does several things. First, it is essential to plant growth. Nutrients move within the soil solution and are absorbed (taken up) from it by plants through the roots (see Chapter 8). Second, it is essential to the microorganisms that live in the soil and decompose organic matter and recycle plant nutrients (see Chapter 4). Third, it is important in the weathering process by accelerating the breakdown of rocks and minerals to form soil and release plant nutrients (*see* Chapter 3).

Water in the soil influences the timing of many farming operations, such as when to till, when to plant, and when and how to apply herbicides and/or fertilizers. Soil water influences the choice of crops to be grown. In areas where rainfall and soil water are sufficient, corn may be grown. In areas with less rainfall and/or more evaporation, there is less soil water available, and grain sorghum (milo) is more likely to be selected by farmers. To effectively manage available water resources, it is important to understand the processes of water movement in soils and uptake by plants.

Hydrologic Cycle

Hydrology is the study of the movement of water on the earth. The hydrologic cycle (Fig. 6.1) is used to summarize all the processes involving water in the environment. When the hydrologic cycle is considered on a global scale, it is common to begin with evaporation of water from the oceans. Evaporation also

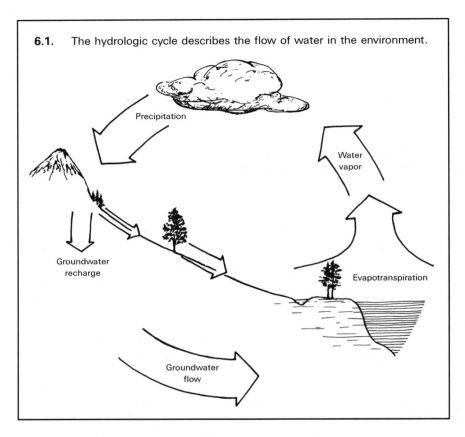

6.1. The hydrologic cycle describes the flow of water in the environment.

Precipitation

Water vapor

Groundwater recharge

Evapotranspiration

Groundwater flow

occurs from the land, and a small amount of water vapor comes directly from snow and ice in alpine and polar regions through *sublimation* (tranformation of ice directly to water vapor). Water vapor in the atmosphere forms clouds and the water falls back to earth in precipitation (rain, snow, sleet, and hail). Precipitation that falls back on the ocean can be evaporated again. Snow that falls in polar or mountainous regions may be stored for decades or centuries before it melts.

Some of the precipitation that falls on land is intercepted by vegetation and evaporates back to the atmosphere but most of it reaches the soil surface. Precipitation that reaches the soil can enter the soil or run off to streams, marshes, or lakes. Surface water eventually evaporates, seeps farther into the earth, or flows back to the oceans, where it can evaporate and start the cycle again. Water that enters the soil is of most importance to plant growth. This water can evaporate from the soil surface, be taken up by plant roots (and evaporate from leaves), or pass through the root zone.

The global hydrologic cycle is very complex and involves processes that occur on large scales (precipitation) and over long periods of time (melting of glaciers). Nonetheless, parts of the hydrologic cylce have strong implications for food and fiber production. For instance, in several areas of the world, water from melting snow and ice is captured in reservoirs and used to irrigate crops sometimes hundreds of miles (kilometers) away.

Soil Water Budget

Although soil water is just one component of the hydrologic cycle, it represents the crucial reservoir of water for the growth of most crops. An easy way to monitor water in soils is to consider the soil's water budget. Just as a person may have a financial budget with inputs (income, investments, and so on) and outputs (food, clothing, shelter, and so on), the soil has a water budget. In agricultural settings there are two inputs: precipitation and irrigation. Water from precipitation and/or irrigation can either move into the soil (***infiltration***), or it can run across the soil surface to a stream, marsh, or lake (***runoff***).

What happens to water after it enters the soil? It can move through the soil and out of the root zone (***percolation***) and eventually to groundwater. It may also evaporate, either directly from the soil or from plant leaves after being taken up by roots. Thus, the soil's water budget has two inputs (precipitation and irrigation) and three outputs (runoff, percolation, and evaporation). Finally, the water budget even has a "savings account," which is the amount of water stored in the soil profile (Fig. 6.2). The water balance can be put in equation form:

$$P + I = Q + E + D + S$$

where P is precipitation, I is irrigation, Q is runoff, E is evaporation, D is percolation, and S is soil water storage. Each term in the water balance equation would have units of depth such as inches (millimeters). Typically, water balance analy-

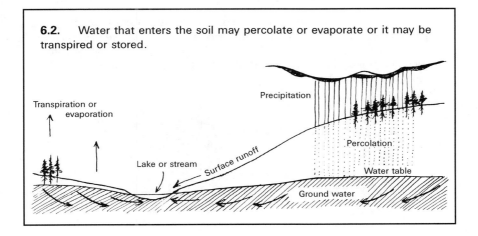

6.2. Water that enters the soil may percolate or evaporate or it may be transpired or stored.

ses are completed for months or years so managers can analyze trends in each term and consider options to optimize water use.

The amount of water that ends up in each term is partly determined by climate and partly by properties of the soil and the requirements of the plants growing in the soil. Of course, humans also have the opportunity to manage the movement of water by choosing which crops to plant, when and how much to irrigate, and types of tillage practices to follow.

Infiltration and Runoff

Precipitation or irrigation that reaches the soil surface is partitioned between infiltration and runoff. The rate of infiltration varies with the texture and physical condition of the soil. Sandy soil, because of its relatively large pore size, has a higher infiltration rate than clay soil with its smaller pore size (Fig. 6.3). If the physical condition of the soil is poor, the infiltration rate is lessened. A sandy soil may have an infiltration rate greater than 1 inch (2.5 cm) per hour, whereas some clayey soils require more than 12 hours for 1 inch (2.5 cm) of water to infiltrate.

Figure 6.4 gives an indication of the rates of runoff and infiltration for a hypothetical rainstorm. The rate of infiltration needs to be known when designing a drainage or irrigation system. One way to estimate infiltration would be to observe how long it takes before water starts to run off (if it ever does) during a rainstorm. If the rate of water infiltration into soil is less than the rate at which rain falls, water accumulates on the soil surface. If enough water accumulates to fill the small depressions, runoff begins. If the amount of rainfall and the duration of the storm are known, then the infiltration rate can be estimated.

Runoff water carries soil with it and erosion occurs. It is generally desirable to hold as much of the rain as possible where it falls to provide water for crops and to protect the soil from erosion. On some soils in humid regions, however, it is

6.3. Soils with large pore spaces, such as sandy soils and well-granulated types, usually have high infiltration and percolation rates, whereas those that have small pore spaces or are in poor physical condition have low infiltration and percolation rates. Runoff occurs if the rate of rainfall exceeds the water infiltration rate.

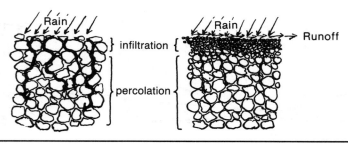

6.4. Runoff and infiltration for a 1.5-inch (38-mm) rainfall in 1 hour. The infiltration rate decreases as the soil wets until runoff begins after 10 minutes. Late in the storm, the runoff and infiltration rates are steady. Runoff would have begun later and been less if the soil had a higher infiltration rate.

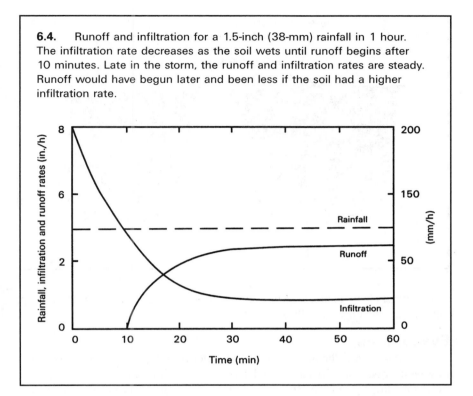

necessary to encourage runoff through a surface drainage system that prevents the soil from becoming waterlogged. It is impossible to avoid all erosion, but it is important for it to be minimized. Soil erosion is discussed further in Chapter 10.

A farmer can influence the infiltration rate of a soil by keeping a protective vegetative cover on the surface and by maintaining good soil structure, both of which help conserve water and soil. By keeping the soil in good physical condition, the topsoil becomes full of "crumbs," which are stable, spongy aggregates into and through which water moves easily. Such aggregates form when plenty of organic matter is present and minimal tillage is used. The farmer who depletes the soil of organic matter by removing crops without returning plant residues or manure is likely to decrease the infiltration rate. When the spongy aggregates do not form, raindrops strike exposed soil, which is beaten to a paste, and a seal forms on the surface. The soil surface then tends to shed water like a roof.

When a sealed-off surface becomes dry, it forms a brittle crust that can inhibit emergence of seedlings. Seedlings such as those of beans and potatoes are strong enough to break through, lifting pieces of the crust like so many trapdoors opened from below. But seedlings of small-seed crops such as oats and even the larger seedlings such as corn may perish without ever emerging (Fig. 6.5). Some-

6.5.   If a plant seedling is not strong enough to lift the soil crust, it dies.

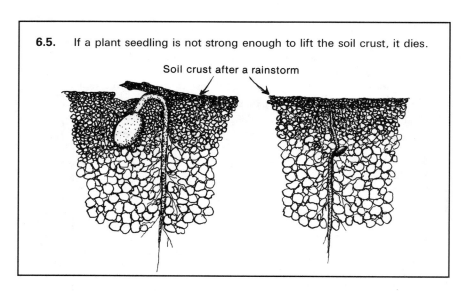

Soil crust after a rainstorm

times farmers have to break up this crust with light tillage after planting the crop, but this may be only partially successful.

Evaporation

Evaporation is the transformation of water from liquid to vapor in response to solar energy, wind movement, and dryness of the air (humidity). Soil water can evaporate directly from the soil or it can be taken up by roots and evaporate from stomates on the leaves of plants. The process of evaporation from stomates is called *transpiration*. Evaporation from soil and transpiration by plants may be combined and called *evapotranspiration* (Fig. 6.6).

6.6.   Soil water returns to the atmosphere by evaporation from the soil surface and by transpiration from plant leaves.

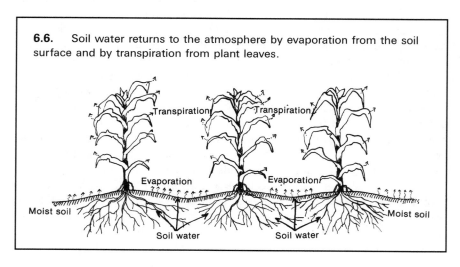

In many parts of the United States, at least half the water lost from farmland is by evaporation. Therefore, farming practices are carried out to reduce this loss and conserve moisture. One effective practice is to leave the dead plants from the previous crop on the soil surface. These crop residues reduce evaporation by shading the soil and blocking water vapor movement.

In small plots of high-value crops, mulches can be used to hold the soil water for plants and thus reduce evaporation (Fig. 6.7). Many kinds of mulches have been used: straw, corncobs, gravel, and more recently, plastic. All have been quite effective. The selection of one mulch over another depends on the specific use and availability of the material. Organic materials such as straw are preferable in situations where the mulch can be incorporated into the soil after each crop. Sand and gravel have the advantage of allowing a higher percentage of small rains to infiltrate into the soil rather than being absorbed by the mulch. This can be an advantage around fruit trees and ornamental plantings.

The use of black plastic as a mulch is increasing in vegetable production because it effectively controls both weeds and evaporation. A variation in this practice is to form plastic or tar paper into a shallow cone around the base of a tree or shrub and cover it with a few inches of gravel. This allows rainwater to enter the soil near the trunk and leaves no place for weeds or grass to grow in hard-to-mow places (Fig. 6.8).

Farmers in dry areas where wheat is a leading crop utilize the principle of mulching by leaving much of the plant residue on the soil surface when tilling after harvest. Creating a dust mulch by frequent tillage of fallow land is now often discouraged because it has been found that little water is conserved by this practice and that soil may be left susceptible to severe wind erosion.

Percolation

In 1856, a French engineer named Henry Darcy was the first to describe how water moves through a saturated soil. He developed his theory by observing the flow through a sand filter used to purify drinking water in the city of Dijon.

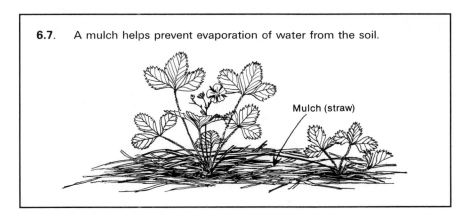

6.7. A mulch helps prevent evaporation of water from the soil.

Mulch (straw)

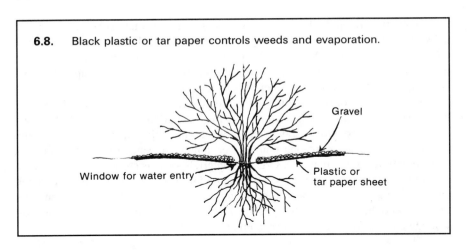

6.8.   Black plastic or tar paper controls weeds and evaporation.

Gravel

Window for water entry

Plastic or
tar paper sheet

The force of gravity causes water to move downward through the soil, particularly in larger pores. Gravitational water percolates until it is adsorbed by drier soil below or it reaches the water table. The *water table* is the level in porous subsurface materials below which all pores are filled with water. This may be within the surface soil, in buried sediments, or in the deep bedrock. *Groundwater* is the water below the water table. The amount of water that percolates through the root zone to groundwater is referred to as *groundwater recharge*. The water table surface has the same general shape as the land surface, but it tends to be smoother. Porous rock layers such as sandstone may be saturated and can be important sources of well water. Such layers are called *aquifers*.

In a flat area, water flow in a uniform soil will be primarily downward. On steep slopes or when there is a gentle slope with a restricting or more conductive layer in the soil (like a layer of clay or gravel), water movement may still be downward but some water may also move downslope or laterally. Lateral water flow in mountainous areas may come to the surface again as discharge from flowing springs. In all cases, plant uptake draws water out of the soil, altering water movement in the soil near the roots.

Wetlands, lakes, and streams in humid regions are often places where the water table comes to the surface and groundwater discharge takes place. Wetlands in drier regions, however, are often groundwater recharge areas, where surface runoff collects and infiltration occurs. Thus, wetlands in these areas serve an important role in replenishing aquifers.

Soil Water Storage

When a soil is wet, gravity is the dominant force in moving water deeper into the soil. At the same time that gravity is pulling water downward, the soil particles are attracting water in all directions by the forces of adhesion and cohesion. ***Adhesion*** is the attraction of a surface for water (for example, the surface of the soil particle), and this force is quite strong. ***Cohesion*** is the attraction of one water molecule for another. The two forces combine so that water is held within small pores between soil particles. Because one of the forces is the attraction a soil particle surface has for water, it follows that a soil with very small particles such as a clay (resulting in more surface area with greater total surface) attracts and holds more water.

When a soil dries out, adhesive forces begin to dominate, and the water remaining in films is held very close to the soil particles so that water movement is very slow. Water movement in unsaturated soils involves a complex combination of gravity and adhesive/cohesive forces. It wasn't until 1907 that an American, Edgar Buckingham, was able to accurately describe water flow in unsaturated soils.

As a plant root absorbs water, it takes some from the film surrounding the adjacent soil particle. Due to cohesion and adhesion, water moves from particles with thicker films to particles with thinner films that are next to the roots. This is called ***capillary movement*** (Fig. 6.9). Compensating water may move to the root from any direction—up, down, or laterally. An important fact is that water moves in the soil toward roots to provide plants with water. Some essential nutrients also can move with the water. Capillary movement, however, is normally very slow in soil, so plants must continually extend their roots into moist pores to prosper. The soil thus has a critical role in agriculture as it acts as a storehouse of water for plants to use until the next rain or irrigation.

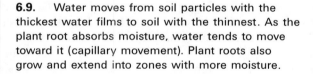

6.9. Water moves from soil particles with the thickest water films to soil with the thinnest. As the plant root absorbs moisture, water tends to move toward it (capillary movement). Plant roots also grow and extend into zones with more moisture.

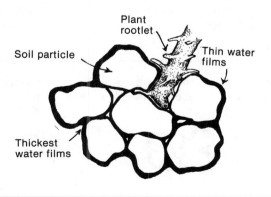

Plant rootlet

Soil particle

Thin water films

Thickest water films

Plant Water Use

Water is essential to all forms of life—both plants and animals. Some plants have low water requirements and are called *xerophytes* (*xero* means little or none and "phytes" comes from the word *phyto*, meaning plant). Some have high water requirements and are called *hydrophytes* (*hydro* means water). Plants with moderate water needs are called *mesophytes* (*meso* means intermediate).

Plants need water to form certain compounds. For example, six parts of water are required for each simple sugar produced. The process of forming simple sugars, called *photosynthesis*, involves the splitting of water (H_2O) into hydrogen and oxygen. The hydrogen combines with carbon dioxide (CO_2) to form sugars, and the oxygen is discharged into the atmosphere through openings called stomates in the leaves of plants.

Much of the water held by the soil is used by a plant sooner or later. The amount held in the soil and the amount available to plants vary with the texture of the soil. As gravity moves water downward, the film of moisture around soil particles thickens and the water in the soil is most readily available to the plant. The maximum amount of water in a soil held against the force of gravity is called the *field capacity*. As water is used by plants or evaporates, the water film around soil particles becomes thinner, is more tightly held by the particles, and is more difficult for the plant to absorb. Eventually, the attraction between the soil and the water is greater than the plant's capacity to absorb it. This amount of water in a soil is called the *wilting point* because the plant can no longer absorb enough water to maintain transpiration and sustain life (Fig. 6.10).

Between these two points, the field capacity and the wilting point, water is available to the plant (Fig. 6.11). The amount that is available varies with soil texture. For example, a sandy soil (which has large particles and low surface area) may hold about 1 inch of water per foot of soil (83 mm/m) and most of the water would be available to plants. A clay soil (which has small particles and high surface area) may hold 4 inches of water per foot of soil (333 mm/m), but because of the strong attraction of clay particles for water, only 1 of these 4 inches may be available. Soils with the greatest amount of available water are usually those with a loamy texture and good structure.

6.10. The water films in A are thickest and the soil is nearly saturated, at B it is about at field capacity, and the thin films in C represent the wilting point.

A   B   C

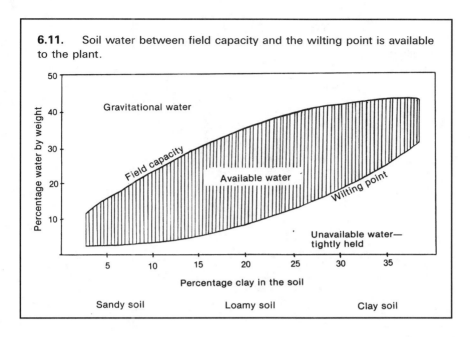

6.11. Soil water between field capacity and the wilting point is available to the plant.

Only a very small percentage of the water taken up and utilized by a plant is for photosynthesis. Water's principal function is to transport nutrients and plant compounds in solution, either upward from the plant roots to the upper leaves or downward into lower leaves or the root system. Most of the water taken up by a plant eventually evaporates at the stomates (Fig.6.12). Evaporation inside the stomates is called transpiration and, especially in hot weather, transpiration helps cool a plant. Less than 1% of the water absorbed by a plant is used in forming plant compounds; the rest is lost via transpiration.

A plant's water use efficiency is determined by measuring the amount of water required to produce a certain weight of dry plant tissue. It takes approximately 500 pounds (225 kg) of water to produce 1 pound (0.45 kg) of wheat (foliage plus grain). Only 5 pounds (2.25 kg), or 1% of this amount actually becomes part of the plant. Alfalfa uses more water, requiring about 850 pounds (385 kg) of water per pound (0.45 kg) of dry matter; while grain sorghum, an efficient water user, may require less than 300 pounds (135 kg) of water per pound (0.45 kg) of dry matter.

Drainage

It is a common occurrence in many regions of the world for the soil to contain too much water during rainy seasons of the year or during winter when evaporation is low. If the soil is waterlogged too long during the growing season, roots die from lack of oxygen or from accumulation of toxic compounds. To rid the soil

6.12.   Water moves into the roots and through the plant primarily by capillary action.

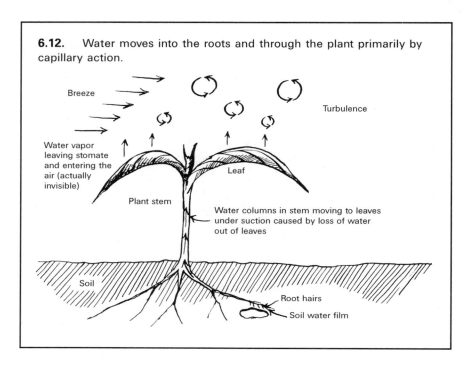

Breeze

Turbulence

Water vapor leaving stomate and entering the air (actually invisible)

Leaf

Plant stem

Water columns in stem moving to leaves under suction caused by loss of water out of leaves

Soil

Root hairs

Soil water film

of excess water, drainage systems have been installed on millions of acres (hectares) of land. Drainage systems can involve subsurface or surface practices or a combination of both. Remarkable increases in crop yields can occur when naturally wet soils are drained.

Subsurface drainage is the practice of burying tubes horizontally in the soil that intercept percolating water and carry it away from the field to some outlet. Drain tubes or tiles were originally made of short sections of concrete or clay tubes, but long lengths of plastic tubing are now more popular. The tile is installed at a depth of about 2 to 6 feet (0.6 to 1.8 m) and has a slight downward gradient to the outlet. Tiles function only when the soil around them is saturated, so that water can flow from the large pores in the soil into the gaps between sections of clay or concrete tile or through holes in the plastic tubing and then out the tile lines (Fig. 6.13). In some cases, a vertical tube is installed from the drain tile to the surface to allow water ponded on the soil surface to enter the drain tile without percolating through the soil. These tile inlets or intakes are commonly found in areas with small depressions that fill with water during heavy rains or spring snowmelt.

Subsurface drain tile can be installed in different patterns, depths, and spacings depending on land slope and location of the outlet (Fig. 6.14A–D). The **random** design (A) is used where there are isolated wet areas. Drain lines are run under each area with perhaps a tile inlet in the larger depressions. **Pattern** (B)

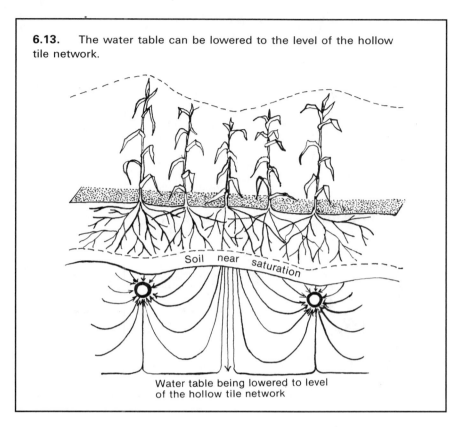

6.13. The water table can be lowered to the level of the hollow tile network.

Soil near saturation

Water table being lowered to level of the hollow tile network

and **herringbone** (C) drainage patterns involve uniform distances between multiple drain lines. The pattern used depends on slope of the land and desired depth of unsaturated soil. An **interceptor** drain (D) is used to intercept lateral flow down a slope that may be creating a wet spot.

Surface drainage involves digging channels in the soil and sometimes also shaping the land surface so water will run over the surface into the channels. Surface drainage is used on soils that have layers with low permeability or in very flat areas like the Red River Valley of North Dakota and Minnesota. In these areas, water either cannot move through the soil fast enough or the slope of the land is too small for subsurface drain tiles to flow effectively. Combination surface/subsurface drainage systems involve subsurface drains using surface drainage channels for their outlets.

To have an outlet for tile drains, some natural stream channels have been straightened and/or deepened (Fig. 6.15). This practice is called *channelization*, and is often criticized for impacting water quality and wildlife conservation. Channelization can be bad when it is used to drain wetlands that ought to be preserved for wildlife habitat or for water quality protection. However, on agricul-

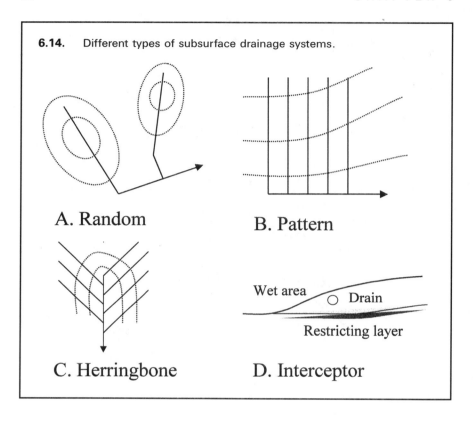

6.14. Different types of subsurface drainage systems.

A. Random

B. Pattern

C. Herringbone

D. Interceptor

Wet area   ○ Drain

Restricting layer

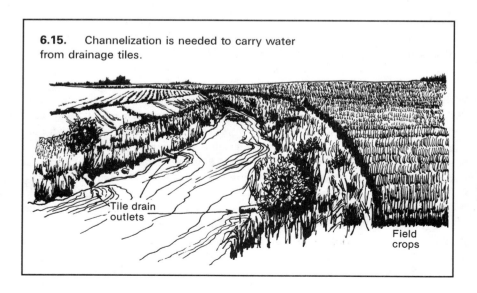

6.15. Channelization is needed to carry water from drainage tiles.

Tile drain outlets

Field crops

tural land that is in crops and pasture, channels can be beneficial to both agriculture and wildlife. When intermittent waterways are deepened, a permanent stream may be formed. As a result, fish thrive, and birds, mammals, and reptiles find an improved environment along the channel banks. Crops also flourish, and they too provide food and shelter for wildlife.

To help offset the loss of wetlands due to farmland drainage and channelization, artificial wetlands can be created. Shallow depressions are made in the soil in low-lying areas and wetland plant species are transplanted into the depression. Water from runoff or drainage tiles is directed into the wetland. The wetland may be designed to be flooded the entire year or only after spring snowmelt and large rainstorms. These constructed wetlands provide wildlife habitat and also may help reduce chemical and sediment transport to streams and increase groundwater recharge.

Irrigation

Since ancient times, various civilizations have utilized water from rivers and wells to ensure more reliable crop production. In the United States, about 16% of the cropland is irrigated. Because yields on irrigated land are usually high, irrigated cropland accounts for a disproportionately large part of crop production. In countries where rice is the staple diet, the people rely on irrigated agriculture for most of their food. Many of these countries have a monsoon climate where the annual rainfall is high but most rain falls during a few months and is followed by an extended period of minimal rainfall. Even in the temperate regions, timely rains can be quite unpredictable from year to year so irrigation agriculture is common.

The first two conditions to consider for irrigation feasibility are a source of high-quality water and adequate soil drainage. The two most common limiting factors in water quality are a high total salt content and a high concentration of sodium ions. In a few cases, other ions such as boron, lithium, and selenium may be present at toxic levels in the proposed water source. On the other hand, if NO_3^- nitrogen is present in irrigation water at high concentrations, such as in the Platte River valley of Nebraska, a credit should be given for the amount of nitrogen applied during irrigation. Sometimes special water-utilization techniques can be employed to enable the use of water of marginal quality.

All irrigation water contains salt, and when irrigation water evaporates, the salts tend to accumulate and might negatively impact plant growth. This is when the internal drainage of the soil becomes important. If salts accumulate, the only way to remove them is by applying more water. When this water moves downward, the salts dissolve and move with the water. This is called *leaching*. Water used for leaching can be either from rainfall or irrigation. The removal of salt from irrigated soil may create other problems. If internal drainage of the soil is good, salts can be leached down through the root zone and either accumulate there or continue to move downward. If soil drainage is poor, salts can move off the soil surface as water moves across it (as in rice production). In some soils with poor internal drainage, tile drainage systems may be installed. As water moves through the soil into the tile drainage system, salts move with it (such as in the Imperial Valley of California). Regardless, the salt concentration in the water increases as it moves across or through the soil. The water coming out may contain 3 to 4 times the salt content compared to the water going into the field.

Discharging water with high salt content into the drainage system of an area can create problems for those downstream from the point of discharge. If salt concentration is too high, it can be detrimental for human and livestock consumption or for reuse as irrigation water for crop production.

If sodium concentration is high in the irrigation water, the soil will develop a high exchangeable sodium level. Sodium is attracted much less to colloidal surfaces than is Ca^{++}, but it takes only about 15% Na^+ on the colloids to cause deflocculation. Soils with high exchangeable sodium will normally develop surface crusts and soil aggregates will disperse. These conditions decrease the rate at which water moves through the soil, which interferes with drainage and salinity control. Scientists at the U.S. Salinity Laboratory developed the *sodium adsorption ratio (SAR)* to characterize the sodium status of irrigation water and soil solutions:

$$SAR = [Na^+]/([Ca^{++} + Mg^{++}]/2)^{1/2}$$

where the concentrations of Na^+, Ca^{++}, and Mg^{++} are expressed as moles of charge/liter. Sodium adsorption ratio values are not used alone but rather with other measures of salinity or sodicity. An SAR below 13 to 15 for a soil extract is generally considered acceptable.

About three-fourths of the irrigation water in the United States comes from surface waters such as rivers and reservoirs where rainwater and/or snowmelt have been impounded. The remaining one-fourth comes from underground aquifers. Some wells tap shallow aquifers that are replenished by precipitation annually, such as in the Central Sands of Wisconsin, but others "mine" deep water that was trapped in aquifers thousands of years ago and is not being replenished to an appreciable degree. This situation, referred to as *overdraft*, is occurring in parts of the High Plains of Texas, where the water table is dropping. This potential problem exists where the Ogallala aquifer, which reaches from South Dakota to Texas, has been supplying irrigation water since the 1930s. The solution for extending the life of the Ogallala aquifer, while maintaining agricultural production, is improved efficiency in the use of irrigation water.

Types of Irrigation Systems

The type of irrigation system used for supplying water to crops is primarily dependent on the topography, the permeability of the soil, and the types of crops being grown.

Furrow irrigation is used extensively for row crops grown on level land or where the slopes are smooth and gentle enough to permit contour furrows the length of the field. The soil must contain enough silt and clay so that the water cannot penetrate so quickly as to enter the soil before it reaches the far end of the furrow. This is a very common system throughout the central and western United States, but it is labor intensive to apply water and maintain the furrows (Fig. 6.16A).

Border irrigation (Fig. 6.16B) involves pumping water to the top of a gentle slope divided by ridges (borders) with a supply point between each set of ridges. It is commonly used where forage is being produced if the lay of the land is suitable.

Basins (Fig. 6.16C) are enclosed by ridges on all four sides so that the land can be periodically flooded. This type of irrigation system works well in the orchards of southern California.

Sprinkler systems (Fig. 6.16D) are growing in popularity because they require the least amount of land preparation and minimal labor. In Wisconsin, there is one farmer who controls 57 center pivot systems from the computer in his office. Many types of sprinkler delivery systems have been used but most modern installations are of the center pivot type. With this system, water is pumped to the middle of the field and forced into a pipe with nozzles that sprinkle the crop as the pipe pivots around the field on wheels driven by electric motors or by water pressure. Most of the units have a radius of 1,320 feet (402 m) that covers a quarter section. Some larger units may have a radius of 2,640 feet (805 m) and are used for an entire square mile (640 acres or 270 ha) of land. An extension can be added that swings out to irrigate the corners of a square field. Plant nutrients (*fertigation*) and agricultural chemicals (*chemigation*) can be metered into the water and delivered to the crops with little additional labor. Sprinkler systems may be used where soils are so sandy and permeable that furrows cannot transfer water to the end of the field.

Trickle (or drip) irrigation (Fig. 6.16E) uses water the most efficiently of all the irrigation systems. It is usually used on small fields of high-value crops. It involves pumping water into plastic tubes with small holes (emitters).The emitters can discharge water either on the surface next to the plants or in the root zone. The water must be of high quality and be filtered to prevent plugging of the emitters. As the cost of irrigation water increases and technology improves, the use of the trickle system will undoubtedly expand.

Subirrigation involves manipulation of a subsurface drainage system to put drainage water back into the root zone. It is presently the least common irrigation system, but it can be used on some alluvial plains and organic soils where the water table can be raised to the level of the plant roots.

Precision irrigation is gaining in popularity where there is significant soil variation within the field, particularly with regard to natural soil moisture conditions. In this system each individual nozzle or emitter can be computer controlled based upon a digitized map. Water with or without fertilizer or other inputs can be adjusted for specific needs of the crop throughout the field.

6.16. Five types of irrigation systems to match the needs of specific crops and soils.

A. Furrow irrigation

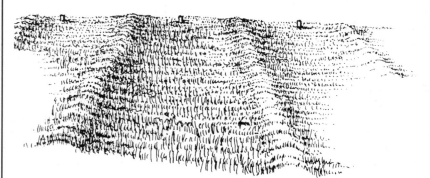

B. Border irrigation

C. Basin irrigation

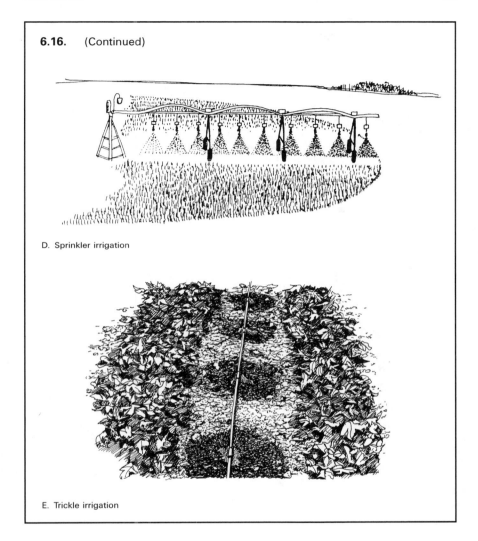

6.16. (Continued)

D. Sprinkler irrigation

E. Trickle irrigation

Water Conservation

The recognition of high-quality water as a valuable resource has led to extensive research on improving irrigation efficiency. Water-conserving techniques now in common use include (1) improved timing of water application based on soil moisture in the root zone (often continuously monitored with sensors buried in the soil), (2) improved application techniques such as trickle or drip irrigation, (3) plastic lining of supply ditches, (4) crop selection, (5) optimized plant population density, and (6) attention to plant nutrition and health maintenance. To minimize water loss by evaporation, plant residues or mulch can be especially effective.

CHAPTER 7

Soil Temperature and Heat Flow

All plants need sunlight to grow. Light from the sun supplies the energy needed for photosynthesis and also warms the soil and air in which crops grow. Soil temperature affects almost every chemical and biological activity that occurs in the soil. Management decisions, such as when to plant, are often based on soil temperature. Knowledge of the flow of energy in the soil-plant-atmosphere system helps us understand how plants respond to the local climate.

Heat Transfer Processes

Energy is defined as the ability to do work. There are three forms of energy; potential, kinetic, and thermal. **Potential energy** is energy that is due to the position of an object. One object placed higher than another of the same mass has greater potential energy. **Kinetic energy** refers to the energy contained in a moving object. If there are two identical cars on the same road and one is going twice as fast as the other, the faster car will have greater kinetic energy. **Thermal energy** relates to the temperature of an object. If you have two identical stones and one has been heated by a fire while the other is frozen, the hot stone will have more thermal energy than the frozen one.

Thermal energy is transferred as a result of a temperature difference within or between objects. Heat always flows from a warm object to a cooler one. Heat can be transferred by conduction, convection, or radiation. All three types of heat transfer take place in the soil-plant-atmosphere system.

Conduction

Heat conduction occurs when kinetic energy is transferred from one molecule to an adjacent, cooler molecule. The ability of a material to conduct heat is called its ***thermal conductivity***. Metals such as copper and iron have high thermal conductivities whereas materials such as wood and plastic have low thermal conductivities and are called insulators. A cooking pan is usually made of metal so the heat from the flame or electric coil is quickly conducted to the food, but the handle is insulated with wood or plastic so the cooks' hands don't get burned.

The thermal conductivity of a soil depends on the proportion of the soil volume occupied by the solid, liquid, and gaseous phases. Most soil minerals have a thermal conductivity about 5 times greater than water, 10 times greater than organic matter, and over 100 times greater than air. When a soil is wet, it has a much higher thermal conductivity than when it is dry because the air in the pore spaces acts as insulation (Fig. 7.1). Nonetheless, as discussed later in this chapter, wet soils are often cold soils because it takes so much energy to raise the temperature of water.

Convection

Heat is transferred by convection when the movement of a heated fluid like air or water is involved. Furnaces that heat buildings by blowing warm air through a system of ducts are sometimes called convection furnaces. Heat can be transferred by convection into soils when warm water from a spring rainstorm infiltrates into a cold or even frozen soil. Convection of heat also occurs at the soil surface when wind blows over a soil that is warmer than the air (Fig. 7.2). Because heat always flows from warmer to cooler objects, heat is transferred from a warm soil to air molecules in a cool wind. This type of convection is called ***forced convection***.

Because warm air is lighter than cold air, heat also can be transferred by convection when warm air rises. This process is called ***free convection***. Ceiling fans in tall rooms are used to blow warm air that rose by free convection back toward the floor. On spring mornings, sunshine on a dark, bare soil heats the soil surface and the air above it, causing the warm air to rise into the atmosphere.

Radiation

The third type of heat transfer is radiation. All objects around us radiate energy in the form of invisible electromagnetic waves. This type of heat transfer

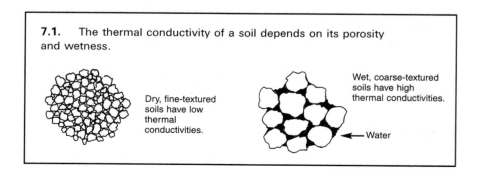

7.1. The thermal conductivity of a soil depends on its porosity and wetness.

Dry, fine-textured soils have low thermal conductivities.

Wet, coarse-textured soils have high thermal conductivities.

Water

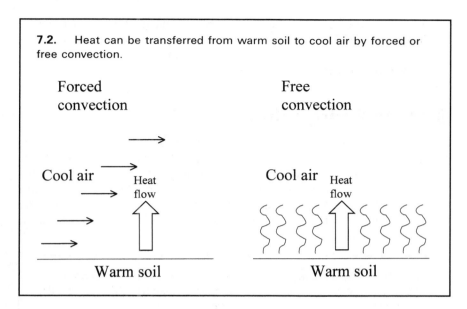

7.2.   Heat can be transferred from warm soil to cool air by forced or free convection.

cannot be felt unless there is a large temperature difference between our skin and the object. If you have a campfire with a ring of rocks around it, you can feel the heat radiating from the hot rocks long after the fire has gone out. Some old buildings and most automobiles also have devices called radiators that are designed to radiate heat to the surrounding air.

The wavelength (distance between two peaks on the wave) of electromagnetic waves emitted by an object depends on its temperature. The surface of the sun, with a temperature of about 10,300°F (5,704°C) radiates energy at a very short wavelength. Thus, sunlight reaching the earth's atmosphere is called ***shortwave or solar radiation***. The planet earth, with an average surface temperature of about 80°F (27°C), radiates energy at a much longer wavelength. Radiation emitted by the earth is called ***longwave radiation*** (Fig. 7.3). Besides the difference in

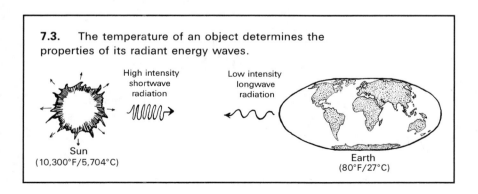

7.3.   The temperature of an object determines the properties of its radiant energy waves.

wavelength, shortwave radiation also transfers more energy (is more intense) than longwave radiation. Plants use solar radiation over only a small range of wavelength. Since this light is so essential to life on earth, it is often measured and is called *photosynthetically active radiation* (PAR).

Surface Energy Budget

In Chapter 6, we described a water budget for a soil. That same idea can be used to describe an energy budget for the soil-plant-atmosphere system (Fig. 7.4). The energy budget is, however, much more complicated than a water budget because heat can flow in both directions (up and down) during different times in the same day and during different seasons of the year. There are also several small terms in the energy budget. For example, photosynthesis is essential to life on our

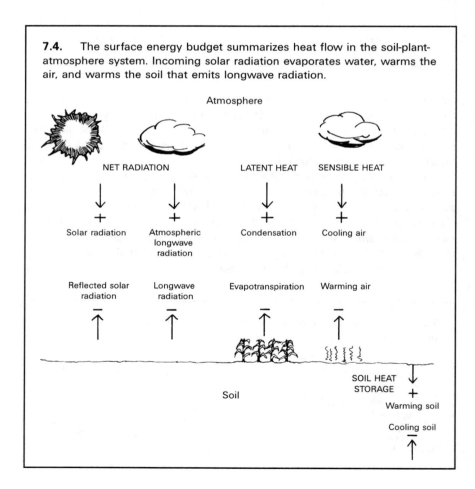

7.4. The surface energy budget summarizes heat flow in the soil-plant-atmosphere system. Incoming solar radiation evaporates water, warms the air, and warms the soil that emits longwave radiation.

planet, but it uses such a small amount of energy that it is often excluded from energy budget analyses.

Here only the four largest terms of the energy budget are considered. Because energy can flow both into and out of the soil, all the terms of the energy budget provide potential inputs or losses of energy. To help follow the flow of energy, a positive sign indicates energy flowing toward the soil and a negative sign indicates energy flowing away from the soil.

Net Radiation

The only significant natural source of energy at the soil surface is sunlight. The first term of the energy budget is called **net radiation (R_n)**. For our purposes, net radiation includes the amount of solar radiation reaching the earth minus the amount reflected by the earth. It also includes the difference between the amount of longwave radiation emitted by the earth toward the atmosphere and the longwave radiation emitted back to the earth's surface by the atmosphere. Usually, net radiation is positive during the daytime (due to large solar radiation input) and negative at night (due to large longwave emission to the atmosphere).

Longwave radiation is trapped in the atmosphere by water vapor and other gases. For this reason, temperatures in desert regions with very hot and dry daytime conditions can be quite cold at night. In more humid areas, higher humidity in the atmosphere helps reduce the contrast between day- and nighttime temperatures. Also, in temperate regions the first killing frost of autumn usually occurs during a night with clear skies because there are no clouds to capture the longwave radiation from the surface.

The amount of longwave radiation absorbed by the atmosphere has implications for global climate as well. Without this natural greenhouse effect, it is estimated that the earth's average temperature would be 0.4°F (18°C) instead of 80°F (27°C). The effect of air pollution on the atmospheric greenhouse effect has led to much concern over accelerated global climate change.

Soil Heat Storage

The next term of the energy budget is **soil heat storage (G)**. During the daytime, the soil surface is warmed by the incoming solar radiation and some of this heat is conducted into the soil. A soil's thermal conductivity affects how much heat is transferred. The heat capacity of a soil also is important. *Heat capacity* is the amount of heat needed to raise the temperature of a substance 1 degree Celsius. Although water has a thermal conductivity about 20 times greater than air, it takes over 3,000 times as much energy to warm an equal volume of water as compared to air. Thus, with the same net radiation and surface temperature, the temperature difference across the surface layer of a wet soil is much less than for a dry soil. Even though a wet soil may have a higher thermal conductivity, it probably conducts less heat because it takes so much more energy to raise its temperature. It was for this reason that wet soils were described as cold soils earlier in this chapter.

At night, the soil surface radiates energy back to the colder atmosphere. As the soil surface cools, heat is conducted upward from deeper soil layers, which results in a reduction in soil heat storage. In the spring, more heat goes into the soil during the daytime than goes out at nighttime, and the soil slowly warms from day to day. In the fall, the opposite happens: less heat is conducted into the soil as the days grow shorter and cooler whereas a greater proportion of heat radiates to the atmosphere during the long nights.

Latent Heat

Some of the energy from net radiation may be used to evaporate water. This term of the energy budget is called **latent heat (*LE*)**. Latent is from the Latin *latere*, meaning hidden or concealed, and is used here because heat is consumed as water changes phases, but there is no change in its temperature.

The *heat of vaporization* is the amount of heat needed to change water from a liquid to a vapor. A relatively large amount of energy is required to change water from its liquid to vapor phase so latent heat may consume a large proportion of the available energy. If the amount of latent heat being consumed is known, the amount of evaporation can be calculated directly. Condensation of water vapor releases energy and thus represents positive latent heat transfer. Condensed water (dew) often forms on crop leaves at night during the summer as the leaves cool and water vapor in the damp air inside the crop canopy condenses. How much dew forms and how long it stays on the leaves are important factors in whether plant diseases occur.

When freezing or thawing of water occurs, another form of latent heat called the *heat of fusion* is involved. Less energy is transferred when water changes between the solid and liquid phases: it takes about one-sixth as much energy to melt ice as it takes to evaporate the same amount of water. Energy is consumed when ice melts and is released when water freezes. Managers of citrus orchards may spray their trees with water before a frost that could damage their fruit. The freezing water warms the leaves and fruit as heat is released as the water freezes.

Sensible Heat

Energy from net radiation also may be used to warm the air layer just above the soil surface. This is called **sensible heat (*H*)** because the warmth of the air can actually be felt. With positive sensible heat, energy from warm air heats the soil; a negative sensible heat means energy is transferred from the warm soil to heat the air.

There is an interaction between the latent and sensible heat transfer. When a soil is wet, the latent heat term is often large because there is a lot of evaporation, and the air stays cool. But when the soil dries, the amount of energy used as latent heat decreases and more energy is used to warm the air and soil. If no rain falls for a long period, the crops may suffer not only from a lack of moisture but also from excessive heat due to increased sensible heat. Understanding sensible heat flow is important because many crops have an air temperature at which they grow best.

Energy Budget Equation

To help understand the flow of energy in the soil-plant-atmosphere system, the energy budget terms can be assigned symbols and put into an equation as was done for the water budget in Chapter 6. Using the sign convention shown in Figure 7.4, the energy budget equation would be

$$R_n - G = -(LE + H)$$

Each term has units of a flux, which means that they represent the amount of energy moving through a unit area in a unit time. The left side of the equation $(R_n - G)$ is called the *available energy* since it represents the amount of energy available for evaporation and heating of the air. The terms on the right side $(LE + H)$ are referred to as the *turbulent fluxes* since wind affects the transfer of water vapor and heat above the soil surface. How the available energy is partitioned between LE and H is frequently of great interest since it relates soil drying with soil temperature. The ratio of LE to H is often calculated and is referred to as the *Bowen ratio*, named for a scientist who developed a theory for estimating evaporation based on energy budget measurements.

Figure 7.5 shows some typical curves for the energy budget terms on an autumn day. The R_n curve is smooth, indicating that there were no clouds. As the

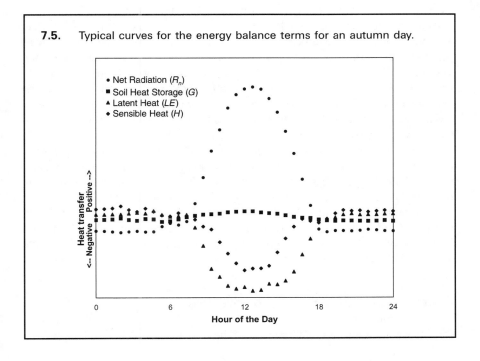

7.5. Typical curves for the energy balance terms for an autumn day.

soil warmed, G became positive during the middle of the day. The soil was moist so LE is more negative than H during the daytime hours, which indicates that evaporation is relatively high. Most of the energy is consumed by LE and H, with little left over for the storage term (G).

If Figure 7.5 had been for a winter day, all of the curves would have been smaller. During winter, the days are short and the sun is low in the sky, so R_n is small. Snow acts as an insulator, so G is small also. Since there is little available energy, both LE and H are small because there is little energy available for evaporation or heating of the air.

Soil Temperature

Soil temperature influences plant growth and crop yields. It also influences date of planting, time to germination, and number of days for a crop to mature. In normal situations, with increased temperatures, root development is faster, availability of nutrients is normally greater, rate of water movement into the soil is usually higher, microbial activity is elevated, and germination and growth of seeds are enhanced.

Ideal soil temperatures for germination vary depending on the crop and the seed characteristics. Most seeds (provided other conditions are ideal, such as adequate soil moisture) require soil temperatures of at least 40°F (4.5°C) to germinate. Cotton, sorghum, and soybean seeds germinate best at soil temperatures around 55° to 60°F (12.7° to 15°C). Corn seed germinates satisfactorily at 50° to 55°F (10° to 12.7°C), and wheat seed germinates at an even lower temperature, around 45°F (7.2°C).

Temperature changes with depth in the soil and varies more at the surface than at deeper levels. At midsummer a soil without vegetative cover may vary in temperature at the surface as much as 40°F (22.2°C) in the course of a day. At a depth of 6 inches (15 cm) in the same soil, the variation in temperature in a day may be only 10°F (5.6°C). At a depth of 24 inches (60 cm) the change in one day would be almost nil (Fig. 7.6). Depth also is a factor in the variation of soil temperature over a year. On an annual basis, the highest temperatures in the upper 1 foot (30 cm) of soil are normally reached in late summer, whereas the lowest temperatures come in late winter. At lower depths of 2 to 4 feet (0.6 to 1.2 m), high and low temperatures lag behind the surface temperatures by 2 to 3 months. The total range in temperatures of soils in a year depends on which part of the world they are located in, but many in the temperate region show a range of as much as 60°F (33.3°C).

In the higher latitudes, this heat gain–loss balance results in the soil being frozen for longer periods of the year. In far northern regions, the subsoil is always frozen. This condition, which is called *permafrost*, exists in much of northern Canada, Alaska, and Siberia.

Factors Affecting Soil Temperature

Soil temperature is influenced by the amount of solar energy that reaches the earth as well as by various soil properties. The amount of heat absorbed depends on

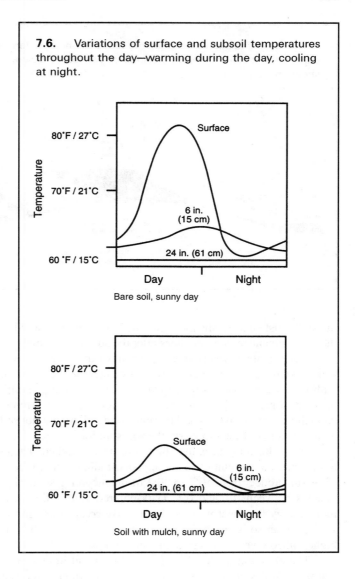

7.6. Variations of surface and subsoil temperatures throughout the day—warming during the day, cooling at night.

the kind and amount of vegetative cover, as well as soil wetness, soil color, the direction of the slope, and the angle at which sunlight strikes the soil surface (Fig. 7.7).

The type or amount of vegetative cover influences how much sunlight reaches the soil surface. A thick crop canopy or layer of mulch shading the soil surface absorbs or reflects much of the incoming solar radiation. In the spring, soils with large amounts of crop residue on the surface warm more slowly than bare soils. Crop residues or other types of mulch also reduce evaporation. Thus, not only do residues shade the soil, but by limiting evaporation, they also keep

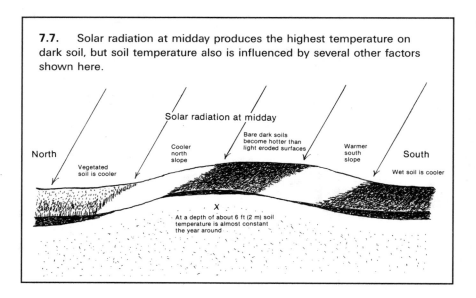

7.7. Solar radiation at midday produces the highest temperature on dark soil, but soil temperature also is influenced by several other factors shown here.

the soil moist so it warms more slowly. These effects should be considered when modifying management practices after the adoption of minimum tillage practices.

Water content affects how fast a soil warms or cools. Wet soils have a higher heat capacity and are much slower to change temperature than dry soils. Soils with high water contents tend to be cold in the spring, resulting in late planting or delayed germination. The ideal condition for soil warming is moist soil with enough water to provide good thermal conductivity but not so much as to require a large amount of heat to warm the soil solution.

Soil color influences the amount of heat absorbed. The term *albedo* is used to describe the amount of incoming solar radiation that is reflected by a surface. Most soils reflect from 10% to 30% of the incoming solar radiation, so they have albedos between 0.1 and 0.3. Dark-colored soils reflect less sunlight, leaving more energy for absorption by the soil. By comparison, most crops reflect more sunlight than soils and have albedos from 0.2 to 0.3. The same soil will have a higher albedo when it is dry than when it is wet.

The direction of the slope (*aspect*) of a soil influences the rate at which it warms. A soil sloping to the south (in the Northern Hemisphere) warms more rapidly in the spring than one sloping to the north. Some specialty crops may be planted on the south side of an east-west row because of the exposure to more direct sunlight. The south side heats relatively rapidly, which speeds germination of seeds by as much as 3 to 5 days.

If the sun is directly overhead, its rays strike the soil surface at right angles and more heat is absorbed than when the sun is at a lesser angle. The sun is more nearly overhead in the summer, resulting in a high level of energy (heat) absorption. In the fall, winter, and spring months the sun appears lower in the sky, and its rays strike the soil surface at a lesser angle, resulting in less heat absorption. These differences are actually due to the absolute surface area over which a given amount of solar radiation is distributed. Obviously, soils that slope toward the sun can intercept more energy and thus be warmer than soils sloping away from the sun.

Managing Soil Temperature

Since temperature has such a profound effect on the rate of chemical and biological processes in the soil, much effort has been put into developing ways to modify the surface energy balance in horticultural and agricultural production. The objectives of these studies have been to optimize the soil thermal regime to enhance plant growth and productivity.

Manipulation of the surface albedo has been used to increase and decrease soil temperature. In hot and arid regions, increasing the surface albedo by whitening the surface with a white powder can double the amount of reflected sunlight. Cooler soil temperatures conserve soil water by decreasing evaporation. Conversely, in subarctic regions, soil temperatures have been increased by blackening crop residue to lower the surface albedo. These types of practices are only temporary and relatively expensive so they are only feasible in extreme cases or with high-value crops.

Slope aspect can also be modified on a small scale to increase soil temperature. This is especially important at planting time at higher latitudes when cold, wet soils delay emergence and early seedling growth. Tillage operations can be used to create a ridge and furrow geometry in east-west rows. Soil in the ridge will warm faster because it dries more quickly and it absorbs more sunlight on its south-facing slope (in the Northern Hemisphere). Seeds planted into the ridge will germinate and grow faster than if they were planted in flat soil.

Mulches, including plastic films, organic byproducts (crop residues, leaves, wood chips, etc.), and gravel, have been used to modify soil temperature. All of these mulches create a water and/or heat barrier at the soil surface. Porous mulches like crop residues and wood chips reduce evaporation by providing a barrier to water vapor. Mulches made of crop residue also have a higher albedo than the underlying soil. Soils under a porous mulch cover will be cooler and moister than unmulched soils during the growing season. Clear plastic film mulches reduce evaporation as well but also increase soil temperature by trapping longwave radiation under the plastic and/or by decreasing the surface albedo. Plastic mulches are common in horticultural applications where young plants require a warm, moist environment.

Combined Effects of Water and Temperature

Soil water and temperature interact with each other in several processes involving soils and plant growth.

Freezing and Thawing

Water moves into the cracks of rocks and minerals. If it freezes, it expands and causes the rocks and minerals to break down. Repeated freezing and thawing eventually results in the formation of soil particles. When tillage creates large soil clods, this same freeze–thaw cycle helps break up the clods over the winter so a smooth seedbed can be prepared in the spring.

In some areas where perennial crops are planted, freezing and thawing in soils causes plants to be lifted out of the soil, a phenomena called *frost heave*. This disrupts the root system. If the action is severe enough, crops such as alfalfa may be destroyed.

Dissolution of Plant Nutrients

Another interaction involves the dissolving of minerals by water. Warm water hastens the dissolution of minerals. Soils in climates where rainfall and soil temperatures are high generally tend to be low in nutrients because the nutrients from dissolved minerals are leached out of the root zone. This is true in the southeastern United States and in many tropical parts of the world.

The water in the soil (the soil solution) also interacts with temperature to increase the availability of nutrients for plant use. High soil temperature and adequate soil moisture mean, in most cases, more nutrients in solution.

Leaching and Eluviation

When water moves through a soil, two other processes take place. One is called *leaching*. As water moves down through the soil, plant nutrients also move downward and, in many cases, entirely out of the zone of root activity so that they become unavailable for use by plants (Fig. 7.8).

The second process is called *eluviation*, the downward movement of very small clay particles from the topsoil to the subsoil. In most cases, the top 6 to 12 inches (15–30 cm) of soil has less clay than soil at a depth of 12 to 24 inches (30–60 cm). This zone of accumulation as a result of eluviation is called the *zone of illuviation* (Fig. 7.8). The effects of leaching and eluviation are greater at higher temperatures.

Plant Species Variation

Slopes that face the sun will tend to be drier and warmer than slopes with a northern aspect (in the Northern Hemisphere). In forested areas of the midwestern United States, north-facing slopes might have a dense oak-hickory-maple forest, while south-facing slopes often have a sparse stand of red cedar and burr oak with grass between the trees. These species are more competitive on the south-facing slopes, which have greater evaporation and higher soil temperatures. Such variations in plant species can affect soil formation. Soils on south-facing slopes will tend to be shallower and contain less organic matter due to the more extreme (warm and dry) climate.

7.8. Water moving through a soil leaches out nutrients and translocates clay particles downward.

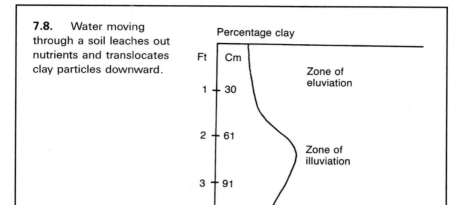

Percentage clay

Ft | Cm

1 — 30

Zone of eluviation

2 — 61

Zone of illuviation

3 — 91

4 — 122

Nutrients and salts may be leached beyond the soil depth

CHAPTER 8

Soil Fertility and Plant Nutrition

Water, carbon dioxide, and certain chemical elements called plant nutrients are essential for plant growth. Water is supplied by either rainfall or irrigation, carbon dioxide from the atmosphere, and the essential plant nutrients from the soil.

Soil Fertility

Soil fertility includes the ability of a soil to hold plant nutrients, the level of plant nutrients present, and the availability of the nutrients for uptake by plants. A soil that has a high level of essential nutrients available for use by plants is usually a productive soil if it also has sufficient soil water and if the crops are well managed. For optimum crop yields and good plant growth, it is desirable to have a high level of the available nutrients.

Plant nutrients exist in the soil in several different forms. They include

1. *Minerals.* Examples include the feldspar group, which is the most abundant group of minerals in the rocks of the earth. Some are high in potassium and others in calcium. Nutrients are released from the minerals by weathering.

2. *Cations or anions that exist on the surface of clay or humus.* These surfaces are called the **exchange complex** and are negatively and positively charged. They attract, hold, and exchange the cations or anions (see Chapter 5).

3. *Chemical compounds that form in the soil under certain conditions.* An example would be the formation of phosphorus complexes on the surface of calcium carbonate.

4. *Soluble ions in the soil solution.* Plants absorb a large portion of their essential nutrients from this source.

5. *Organic matter.* Plant residues and organic matter contain the elements that plants require for growth. As decomposition occurs, these nutrients are released and can be used by plants.

The availability of nutrients to be absorbed by plants varies according to the form in which they exist (Fig. 8.1). Nutrients in soil solution are quite readily available for use by plants. Those on the exchange complex are generally available for absorption by plants but not quite as readily available as those in the soil solution. Nutrients present as complex chemical compounds or as precipitated salts as well as those in organic matter are normally only moderately available depending to a great extent on soil water content, soil temperature, and soil pH. Those present as minerals are slowly available. They are released only as the mineral breaks down during the process of weathering (see Chapter 2 for details on weathering). Most nutrients can be classified as readily available, moderately available, or slowly available. It is desirable to have an adequate supply of nutrients in a readily available form.

Conditions Affecting Level and Availability of Plant Nutrients

Certain soil characteristics influence the availability of nutrients. One is soil texture. Because clay particles provide a part of the exchange complex, the percentage of clay in the soil influences the capability of a soil to hold nutrients. The percentage of clay determines the size of the "nutrient warehouse."

The type of clay also is important. As explained in Chapter 5, three of the types of clay in soils in the United States are kaolinite, illite, and smectite (montmorillonite). Each type has a different capacity to hold nutrients (cation exchange capacity, CEC). In general, the higher the CEC, the more fertile a soil is because more nutrient cations can be held in the soil. The CEC is relatively low for kaolinite (3 to 15 cmol$_c$/kg), moderate for illite (10 to 40 cmol$_c$/kg), and high for smectite (80 to 100 cmol$_c$/kg). It follows that a soil with 20% clay as smectite would have a much greater capacity to hold nutrients than a soil with 20% clay as kaolinite. The clay content and the type of clay are both important in soil fertility.

8.1. Phosphorus exists originally as a complex mineral with very low solubility. Weathering breaks it down into less complex forms, some of which can be used by plants.

	Name	Chemical form	Complexity	Availability of phosphorus to plants
↑ Weathering ↓	Hydroxyapatite	$Ca_{10}(PO_4)_6(OH)_2$	Highly	Very low
	Dicalcium phosphate	$Ca_2(HPO_4)_2$	Moderate	Moderate
	Monophosphate ion	$H_2PO_4^-$	Simple	Readily

Organic matter is another important soil characteristic that, if high enough in content, can favorably impact the availability of nutrients. It has a threefold effect on fertility. The fraction of organic matter that is humus (the colloidal fraction) is similar to clay particles in that it has an exchange capacity ranging from 50 to 200 cmol$_c$/kg (depending on the pH of the soil) and attracts and holds nutrients for plant uptake. As organic matter decomposes, the essential plant nutrients it contains are released and organic acids are formed that increase the availability of most nutrients. An adequate level of organic matter in a soil is generally desirable not only from a plant nutrient standpoint but also because of its favorable effect on soil characteristics such as physical condition, water-holding capacity, and infiltration rate.

Soil water content also influences nutrient availability. Most nutrients utilized by plants are absorbed from the soil solution. A higher level of soil water normally means a higher level of most nutrients in solution for the plants to use.

Higher soil temperature usually leads to greater availability of most plant nutrients, as explained in Chapter 7.

Soil pH, which is a measure of the degree of soil acidity or alkalinity, also influences the availability of nutrients. Most nutrients are at their highest level of availability when the pH is slightly acid to neutral. As a soil becomes more acid, certain nutrients become less available; if a soil becomes alkaline, the availability of certain nutrients decreases.

Plant Nutrition

Essential Elements

At least 16 elements, called plant nutrients, are essential for plant growth (Table 8.1). The first group includes three elements—carbon, hydrogen, and oxygen—which are the basic building blocks of all plant compounds. The initial product of photosynthesis is the simple sugar $C_6H_{12}O_6$. The carbon and oxygen come from carbon dioxide, and the hydrogen comes from water. The oxygen in the water is given off by plants and goes back into the atmosphere. This process assures us of a continuing source of oxygen.

The second group of essential elements, called *macronutrients*, consists of nitrogen, phosphorus, potassium, sulfur, calcium, and magnesium. They are

Table 8.1.   Elements required for plant growth

Basic Nutrients	Macronutrients	Micronutrients
Carbon (C)	Nitrogen (N)	Iron (Fe)
Hydrogen (H)	Phosphorus (P)	Zinc (Zn)
Oxygen (O)	Potassium (K)	Manganese (Mn)
	Sulfur (S)	Copper (Cu)
	Calcium (Ca)	Boron (B)
	Magnesium (Mg)	Molybdenum (Mo)
		Chlorine (Cl)

classified as macronutrients because they are used in relatively large quantities by plants.

Another group of seven elements is called *micronutrients* because they are normally used in smaller quantities. This group includes iron, zinc, manganese, copper, boron, molybdenum, and chlorine.

Some scientists contend that some other elements also may be essential for plant growth. Included in this group are silicon and sodium. These two elements plus vanadium, cobalt, and iodine are often called *beneficial element*s because they can be used by plants as substitutes for nutrients that are essential.

Approximately 90% of the dry weight of a plant is made up of carbon, hydrogen, and oxygen; the balance consists of the other essential elements. Most of this remainder consists of the elements classified as macronutrients, whereas less than approximately one-tenth of this 10% is in the micronutrient group. Several elements not known to be essential also may be included in the 10%.

Natural Sources of Plant Nutrients

Carbon, hydrogen, and oxygen are supplied by carbon dioxide and water. The remaining 13 elements are normally absorbed in ionic form from the soil (Fig. 8.2), although small amounts can be taken in through the leaves if placed there in solution by precipitation, foliar application, or sprinkler irrigation

Nitrogen comes originally from the atmosphere, which is nearly 79% nitrogen, and is in a form that plants cannot use. Bacteria and leguminous plants join together in a process called *symbiotic nitrogen fixation*, which, when combined with other steps in the nitrogen cycle, provides nitrogen in a form usable by plants (see Chapter 4). Nature has other ways of converting atmospheric nitrogen to a form usable by plants. They include fixation of nitrogen by soil bacteria without the help of legumes and the action of lightning discharging in the atmosphere, causing nitrate to be formed, which is then brought to earth by rain. These latter two sources of nitrogen provide relatively small quantities for plant growth.

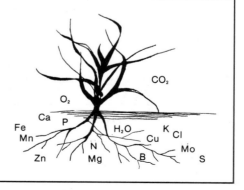

8.2. Carbon and oxygen come from carbon dioxide in the air, hydrogen from water in the soil, and the other elements are absorbed by plants from the soil.

The remaining 12 essential elements are naturally derived from the weathering of rocks and minerals of the earth. Sulfur is often emitted into the atmosphere by coal-burning facilities as sulfur dioxide, which is moved by air currents and then carried to the soil by precipitation (Fig. 8.3). This can also be true for a small portion of the nitrogen used by plants. Phosphorus, for example, comes principally from a mineral called apatite, whereas magnesium comes from minerals such as serpentine and dolomite. When rocks and minerals undergo weathering, elements are released and become part of the soil system. Plant nutrients also are added through irrigation water, but they originally came from minerals.

Role of Essential Plant Nutrients

Each plant nutrient plays one or more special roles in plant growth. A nutrient may be the essential part of a plant compound, thus providing it with a structural base. Calcium, for example, is part of calcium pectate, which is a compound that is a part of the plant cell wall.

Other nutrients may be essential for making compounds involved in plant growth processes, such as phosphorus as a part of adenosine diphosphate and adenosine triphosphate, which are two compounds involved in the transfer of energy within a plant. The compounds used for storage of plant foods such as protein require nitrogen and sulfur.

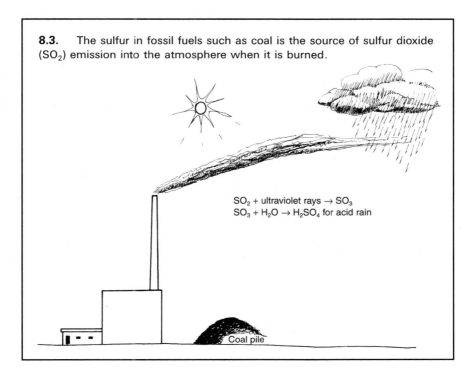

8.3. The sulfur in fossil fuels such as coal is the source of sulfur dioxide (SO_2) emission into the atmosphere when it is burned.

SO_2 + ultraviolet rays $\rightarrow SO_3$
$SO_3 + H_2O \rightarrow H_2SO_4$ for acid rain

Coal pile

Another group is involved in the regulation of certain enzymatic processes. Enzymes in plants, with names such as catalase and lactase, act as catalysts or activators. They often contain micronutrients such as iron and copper. Table 8.2 lists each nutrient, one of its functions in plant growth, and some deficiency symptoms.

Determining Nutrient Need

To be able to produce top yields, all essential plant nutrients must be present in adequate quantities. The most common element to be deficient for most crops and lawns is nitrogen. Phosphorus is normally the second most common element to be deficient. Potassium, calcium, and magnesium are often lacking in soils in the eastern half of the United States where rainfall normally exceeds 25 inches (635 mm) per year; whereas in the western United States, these elements are well supplied in most soils.

Several methods are available to determine if a nutrient is deficient and the quantity needed to relieve the deficiency. These methods include chemical analyses of the soil and the plant, nutrient deficiency symptoms, and growth tests.

Chemical Analysis of the Soil

The method most often used for determining nutrient need is chemical analysis of soil. To determine need most accurately, two things are required:

Table 8.2. Essential plant nutrients, function in plant growth, and deficiency symptoms

Plant nutrient	Function in plant	Deficiency symptom
Nitrogen	Essential part of amino acids, protein, and chlorophyll	Yellowing of midrib of lower leaves
Phosphorus	Part of energy transfer compounds	Reddish-purple color of leaves of young plant
Potassium	Regulation of osmosis and water use and transportation system	Browning of outer edges of lower leaves
Calcium	Formation of calcium pectate used in cell walls	No development of terminal buds and apical tips of roots
Magnesium	Central atom of the chlorophyll molecule	Interveinal chlorosis of middle or lower leaves
Sulfur	Essential part of three amino acids essential for protein formation	Uniformly chlorotic upper leaves and slow growth
Iron	Component of chlorophyll and cofactor for enzymatic reactions	Interveinal chlorosis in young leaves
Zinc	Involvement in auxin metabolism and part of dehydrogenase enzyme	Spotted white or yellow areas between veins of upper third of leaves; also lack of terminal growth
Manganese	Electron transport and part of enzyme system	Interveinal chlorosis in young leaves
Copper	Part of oxidase enzyme system	Yellowing and stunting of young leaves
Molybdenum	Part of nitrate reductase enzyme	Yellowing of midrib of lower leaves
Boron	Growth and development of a new meristematic cells	Pale green young leaves; leaves die and terminal growth ceases
Chlorine	Osmotic and cation neutralization	Partial wilting and loss of leaf turgor when moisture is adequate

(1) a soil sample that truly represents the field in question, and (2) the chemical method that has been adequately researched and properly correlated/calibrated for the crops and soils in question.

Soil Sampling. There are two recommended approaches in taking soil samples from farm fields. They are sampling (1) by soil type or (2) on a grid basis. The same basic principles apply to sampling lawns and gardens but on a smaller scale.

To sample by soil type, diagram a field by soil type (such maps are available in soil surveys) and obtain a composite sample from each soil type (see Fig. 8.4). For each composite sample, take 10 to 15 individual cores of soil from one soil type, mix them together thoroughly, and remove approximately 1 pint (0.5 L) or 1 pound (0.4 kg) to submit for testing. Repeat the same process for each soil type in the field. If one soil type area in a field is too small to be fertilized separately (i.e., less than 5 acres, or around 2 ha, in size), do not sample any of the area.

To sample on a grid basis, divide a field into 3- to 5-acre (1.2- to 2.0-ha) squares as a grid. Take one composite sample (of 8 single-sample core samples) from each square. (See Fig. 8.5.)

Composite samples should always be placed in a clean container to avoid contamination. It is best to use containers provided by the soil testing laboratory, if available. Be sure to provide the producer's name, address, and field number for each sample. Also provide information on cropping history for at least two years, previous fertilizer use and manure applications, and yield levels, and anticipate yield potential for the next crop. Be sure to choose a well-qualified lab whose agronomist is familiar with the soils and crops in the producer's locale. State-operated soil testing laboratories vary in their specific instructions for soil sampling so it is advisable to check with the local county agricultural extension agent.

For grid sampling, various patterns of sampling can be used by shifting from the center of the grid to randomize the sites. This approach to sampling is being used for the computerized application of nutrients or variable-rate application, often called precision or site-specific nutrient management programs.

Methods of Testing. Many methods of soil testing are available. Some measure the total content of a nutrient in the soil, while other tests attempt to measure the "available" nutrient levels. Most testing that is done to predict fertilizer needs for a crop is in the second category, which is to provide an index, or an estimate, of the nutrient-supplying ability of the soil. For any method to work, the soil tests have to be evaluated on the basis of actual crop response. Each test must be correlated/calibrated with field experiments and fertilizer trials. Soils that contain a high amount of available nutrients require less fertilizer input than do soils that contain a low amount of available nutrients. The test is then calibrated to determine the amount of each nutrient needed to maximize profit from fertilizer application. By using the best soil-testing procedures and sound fertilizer recom-

8.4.   Proper collection of soil samples is extremely important. Tests made on carelessly taken samples can be misleading and costly.

Step 1

Take one sample from each major soil type in the field.

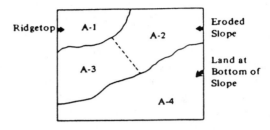

Step 2

Take the sample from all over the area with a subsample from 10–15 places.

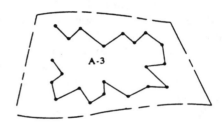

Step 3

Use an auger, a probe, or a spade; scrape the surface inch to the side; and sample to plow-depth.

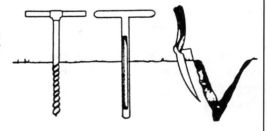

Step 4

Use a clean box for the sample; fill out the information sheet; and send to the lab.

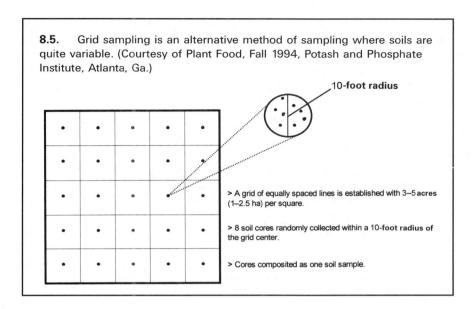

8.5. Grid sampling is an alternative method of sampling where soils are quite variable. (Courtesy of Plant Food, Fall 1994, Potash and Phosphate Institute, Atlanta, Ga.)

10-foot radius

> A grid of equally spaced lines is established with 3–5 acres (1–2.5 ha) per square.

> 8 soil cores randomly collected within a 10-foot radius of the grid center.

> Cores composited as one soil sample.

mendations based on adequate field research, laboratories should be able to predict the optimum economic rate of fertilizer that is environmentally sound.

Potential yields vary; hence, the relationship between soil test values and crop response will vary. This is because yield is affected by climate, disease, and weeds as well as soil fertility. The interpreter of the soil test results should also consider (and should be knowledgeable about) potential yield levels for any given area or even specific farms if possible. Previous yield history, cropping systems, and fertilizer practices, if known, can be most helpful.

Chemical Analysis of the Plant

The procedure for collecting plant samples for chemical analysis is similar to collecting soil samples. First, determine whether the crop in a field is relatively uniform. Then select a plant part (or parts) from about 15 places in the field. The plant part to be sampled depends on many factors—age of plant, type of plant, nutrient to be tested, and so on. Contact your local farm advisor or laboratory consultant for information on how to sample a specific crop.

A report from the testing laboratory provides results on the soil and/or plant tests and normally gives recommendations on the type and quantity of nutrients that need to be added.

Nutrient Deficiency Symptoms

Nutrient deficiency symptoms in plants may be seen as poor growth, lack of green color (chlorosis), or browning of tissue (necrosis). The best-known nutrient deficiency is the one caused by lack of nitrogen. On a corn plant, for example, the green tissue along the midrib of the lower leaves turns yellow. The type of symptom and its location on the plant suggest the nutrient that is deficient. Deficiency symptoms for some nutrients may show up on older, mature leaves or on the younger, new growth. See Figure 8.6 for common deficiency symptoms for various nutrients. Table 8.2 also lists common deficiency symptoms for each plant nutrient.

Biological Growth Tests

Biological growth tests also may be used to determine nutrient needs. A simple method is to split a field and apply one type of nutrient on one half and another nutrient on the other half. Or try one rate of a nutrient on one half and double the rate on the other.

Biological tests can also be used in greenhouses for a short growth period. For such "pot" tests, up to a gallon (4 L) of soil is brought into the greenhouse and divided into small, one-pint (0.5 L) containers, which receive various rates of the nutrients in question (leaving one untreated). Rapid-growing plants (such as small grains) are planted, harvested in a short time (such as 30 days), and

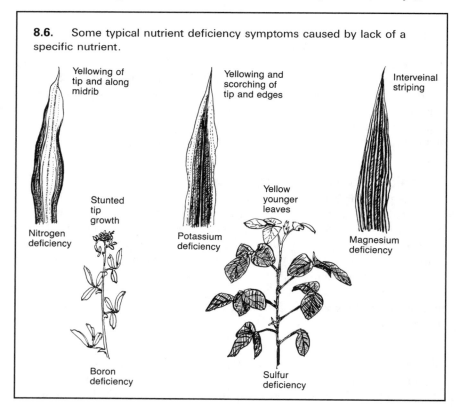

8.6.   Some typical nutrient deficiency symptoms caused by lack of a specific nutrient.

Yellowing of tip and along midrib

Yellowing and scorching of tip and edges

Interveinal striping

Stunted tip growth

Yellow younger leaves

Nitrogen deficiency

Potassium deficiency

Magnesium deficiency

Boron deficiency

Sulfur deficiency

weighed to determine which rate provided the greatest growth.

Adding Plant Nutrients

If the level of essential plant nutrients in the soil is low or their availability is decreased for some reason, nutrients need to be added to achieve good crop yields.

Fertilizers

Each essential plant nutrient (except carbon, hydrogen, and oxygen) can be applied as a commercial fertilizer. There are many different types and forms—so many, in fact, that it would be difficult to describe all of them here. But a few are discussed below.

Fertilizers come in dry, liquid, and gaseous forms. Some contain only one essential nutrient, whereas others contain two or more. Percentages of the essential nutrients in a fertilizer also vary widely. The percentage of a nutrient in a fertilizer is important because it determines the amount to use per acre for a given quantity of a needed nutrient. The percentage of a nutrient or nutrients in a fertilizer is guaranteed to be at a minimum level or above as required by state laws.

A nutrient guarantee is expressed in three numbers, such as 20-10-5. The first is the percent of nitrogen (as N), then phosphorus (as P_2O_5), followed by potassium (as K_2O). Other examples of grades are given in Table 8.3. If there are other guaranteed nutrients present, they are listed as additional numbers with the symbols for the elements. For example, if the 20-10-5 given above also contains 2% zinc and 1% manganese, the grade would show as 20-10-5 + 2% Zn + 1% Mn. This grade guarantee is always listed on the fertilizer container (whether it is in a sack, box, or bottle) and also on the invoice. The manufacturers of the fertilizer make the guarantee. This assures the producer of purchasing the correct product.

A fertilizer that contains only one nutrient is called a *straight fertilizer* (or a fertilizer material). An example would be ammonium nitrate, which contains only nitrogen as a nutrient. *Mixed fertilizers* containing two or more nutrients would be a mixture of two or more straight fertilizers. An example would be mixing ammonium nitrate (33-0-0) with a calcium phosphate (0-46-0) to produce a grade of fertilizer that might be 16-20-0.

Other terms often used to describe fertilizers are complete and balanced. A *complete fertilizer*, a term of little significance from a crop production standpoint, is one that contains all three of the primary nutrients and would be a grade such as 24-10-8 (Fig. 8.7). A *balanced fertilizer* may contain only one essential nutrient or sev-

Table 8.3. Fertilizer grades

Grade	N	P_2O_5	K_2O
	(%)	(%)	(%)
20-10-5[a]	20	10	5
0-20-20[a]	0	20	20
10-30-10[a]	10	30	10
0-46-0[b]	0	46	0
33-0-0[b]	33	0	0

[a] Mixed fertilizers with two or more nutrients.

[b] Straight fertilizers with only one nutrient.

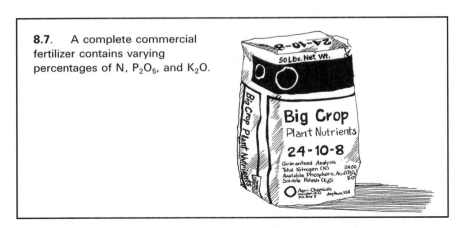

8.7.   A complete commercial fertilizer contains varying percentages of N, P$_2$O$_5$, and K$_2$O.

eral. A fertilizer is balanced if it is in the correct ratio to meet the needs of the crop and correct the deficiencies in the soil. A straight nitrogen fertilizer such as urea (45-0-0), for example, would be a balanced fertilizer if only nitrogen is needed. Or if nitrogen, phosphorus, potassium, zinc, and boron are needed, the balanced fertilizer would contain all elements, for example, 15-10-5 + 1% Zn + 0.5% B.

Nitrogen, the fertilizer nutrient used in greatest quantities, comes in all three forms. Anhydrous ammonia (NH$_3$) is the principal source of nitrogen used in the United States. It is in a gaseous form when applied to the soil (Fig. 8.8) but is stored as a liquid when under pressure or at low temperatures. It contains only one nutrient and its nitrogen content is 82%. This is the highest nutrient concentration in any commonly used fertilizer. Ammonia is the base for producing other nitrogen fertilizers, such as the examples given in Table 8.4.

After ammonia is manufactured, it is mixed with various acids in liquid form and the resulting product is then dried to the solid form (Fig. 8.9). While in the liquid state, urea and ammonium nitrate are often mixed to form a solution containing 28% or 32% nitrogen. This is the second most widely used source of nitrogen in the United States (Fig. 8.10).

Phosphorus fertilizers are derived from a mineral called **apatite**, which is a calcium phosphate and is a form in which the phosphorus is not readily usable by

Table   8.4.   Combinations used to produce nitrogen fertilizers

Combinations	Product	Percentages (N-P$_2$O$_5$-K$_2$O)
NH$_3$ + HNO$_3$	NH$_4$NO$_3$ (ammonium nitrate)	33.5-0-0
NH$_3$ + H$_2$SO$_4$	(NH$_4$)$_2$SO$_4$ (ammonium sulfate)	21-0-0
NH$_3$ + H$_3$PO$_4$	NH$_4$H$_2$PO$_4$ (ammonium phosphate)	11-48-0
NH$_3$ + CO$_2$	(NH$_2$)$_2$CO (urea)	45-0-0

Note: NH$_3$ = ammonia, HNO$_3$ = nitric acid, H$_2$SO$_4$ = sulfuric acid, H$_3$PO$_4$ = phosphoric acid, CO$_2$ = carbon dioxide.

8.8. Nitrogen may be applied as anhydrous ammonia (NH_3) gas fed from a pressure tank through hollow knives that cut into the soil.

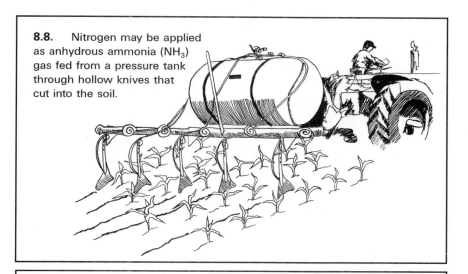

8.9. Most nitrogen fertilizers start with ammonia, which reacts with various acids. They exist in gaseous, dry, or liquid forms.

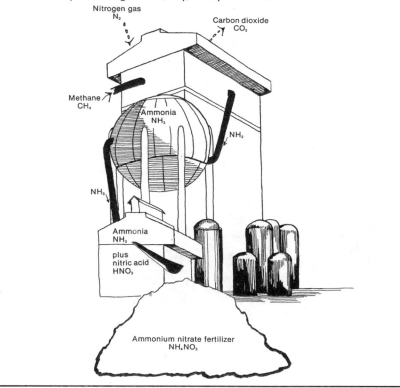

Nitrogen gas
N_2

Carbon dioxide
CO_2

Methane
CH_4

Ammonia
NH_3

NH_3

NH_3

Ammonia
NH_3
plus
nitric acid
HNO_3

Ammonium nitrate fertilizer
NH_4NO_3

8.10.   Liquid fertilizer may be applied to the soil or, if sufficiently diluted, it can be used as a foliar application.

plants. The mineral is mined from deposits just below the surface of the soil in Florida and Idaho in the United States and in Morocco and the former Soviet Union.

Apatite is commonly called rock phosphate (Fig. 8.11). It is treated with an acid (either sulfuric or phosphoric) to produce a calcium phosphate (either 0-20-0 or 0-46-0) in which the phosphorus is in a more usable form than in apatite. Phosphoric acid also can be produced from the apatite and can then be treated with ammonia to produce an ammonium phosphate (Table 8.4). Phosphorus fer-

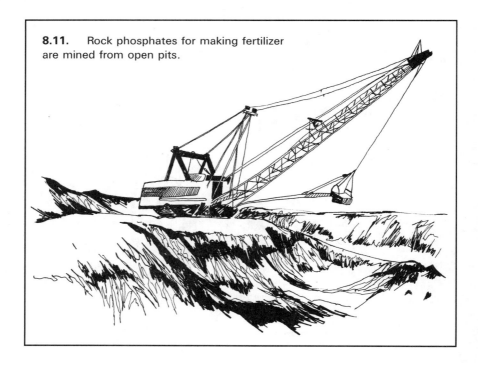

8.11.   Rock phosphates for making fertilizer are mined from open pits.

tilizers are available either in the liquid or dry form. Phosphorus is second to nitrogen in quantity used in the United States.

Potassium fertilizers are manufactured from minerals such as muriate of potash or langbeinite, which occur in deposits in the earth. Some deposits such as those in Canada are fairly shallow, whereas others such as those in New Mexico are quite deep (Fig. 8.12). Muriate of potash is refined to produce potassium chloride (0-0-60); langbeinite is used to produce potassium-magnesium sulfate (0-0-22 + 11% Mg). Other common potassium fertilizers are potassium sulfate (0-0-50), potassium nitrate (13-0-44), and potassium phosphate (0-26-26).

The other fertilizer elements come from various sources. Calcium, for example, comes mostly from limestone (calcite) and gypsum. Magnesium is used either as potassium-magnesium sulfate, limestone (dolomite), or magnesium sulfate. Sulfur is usually applied as elemental sulfur, a thiosulfate, or as one of the sulfate forms such as ammonium or potassium sulfate.

The micronutrients iron, zinc, manganese, and copper are usually used in one of three forms. A **sulfate salt** such as zinc sulfate is common. Another is called a **chelate**, which is an organic form that reacts with the micronutrient to make a relatively soluble product. A third form is an **oxide**, such as zinc oxide.

It is important to use the source of fertilizer best suited for any given crop and condition. A local fertilizer dealer, consultant, or agricultural agent should be consulted for specifics on the best one to use.

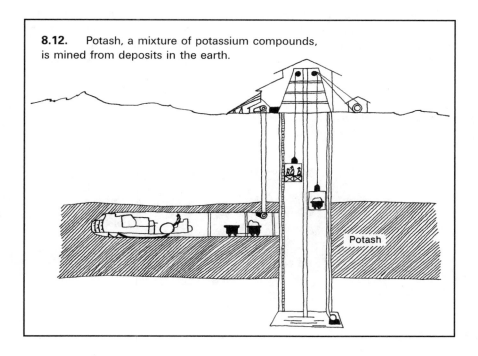

8.12.   Potash, a mixture of potassium compounds, is mined from deposits in the earth.

Potash

Soil Amendments

A soil amendment is a material added to change the soil in some manner—either chemically, physically, or both. Examples include limestone, which is used primarily to increase the pH of a soil and make it less acid. Sulfur is often used to decrease pH and make it less alkaline. Calcium sulfate (gypsum) is used as an amendment on soils with too much sodium, which causes a poor physical condition. The calcium in the gypsum replaces the sodium, which combines with the sulfate thus improving physical condition if the sodium sulfate is leached from the soil.

These soil amendments also provide plant nutrients if they are needed. For example, limestone provides calcium and in some cases magnesium. Gypsum provides calcium and sulfur.

Animal Manure and Green Manure Crops

Animal manures are excellent sources of plant nutrients. Because manure is a residue of unused plant material, it is similar to organic matter and therefore contains all the essential nutrients. When manure decomposes in the soil, its nutrients are released and made available for plant uptake.

Manure not only serves as a source of plant nutrients but also adds organic matter to the soil, which improves its physical condition, water-holding capacity, cation exchange capacity, and similar desirable properties. For maximum value, manure should be injected or worked into the soil as soon after application as possible (Fig. 8.13). A disadvantage of manure is that it often contains weed seed.

The nutrient and moisture content of manure is quite variable, depending principally on the types of feed utilized by the animals and how the manure is handled before it is applied on the field. It should be applied as often as conve-

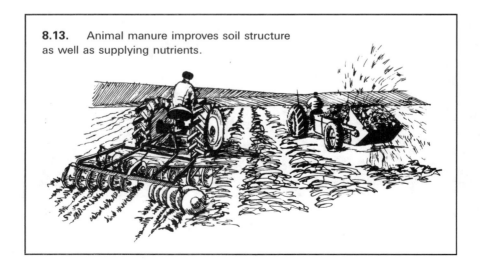

8.13. Animal manure improves soil structure as well as supplying nutrients.

nient to do so, but in most cases, manure has to be stored and applied later. Alternate wetting and drying in a pile results in the release of ammonia gas into the atmosphere as the manure dries. When rewetted by rain, nitrates leach out and may present a danger if allowed to run off into a water system being used by humans or animals. Many states have laws that restrict the runoff from feedlots and manure piles and require owners to provide holding ponds or lagoons (Fig. 8.14). If a manure pile dries without subsequent rewetting (in dry areas), loss of ammonia is minimized.

The average nutrient value of manure from beef feedlots in Texas is shown in Table 8.5. The nutrient content of manure also is quite variable, thus, only averages are presented in the table. The values on moisture content range from a low of 8% to a high of 62%, with an average of 33%. One ton (900 kg) of manure with a moisture content of 33% and average nutrient percentages would be equivalent to 400 pounds (180 kg) of 8-6-11 grade fertilizer.

When animals are confined in barns most of the time, their manure is commonly stored for six months or more in lagoons or large tanks and the moisture content is more likely to be around 90%. It is injected into the soil as a slurry from

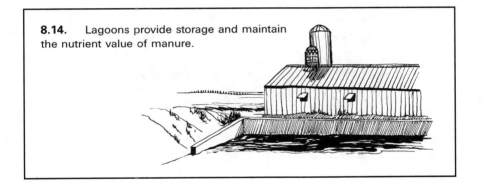

8.14. Lagoons provide storage and maintain the nutrient value of manure.

Table 8.5. Average content of essential elements in beef feedlot manure, based on 30% moisture content

Nutrients	Content	Pounds/ton (kg/metric ton) of manure
	(%)	
Nitrogen (N)	1.6	32 (16)
Phosphorus (P_2O_5)	1.3	26 (13)
Potassium (K_2O)	2.2	44 (22)
Calcium (Ca)	0.7	14 (7)
Magnesium (Mg)	0.2	4 (2)
	(ppm)[a]	
Iron (Fe)	1525	3.00 (1.50)
Zinc (Zn)	100	0.20 (0.10)
Manganese (Mn)	105	0.21 (0.105)
Copper (Cu)	7	0.02 (0.01)
Boron (B)	15	0.03 (0.015)

[a] ppm = parts per million or 1/10,000 of a percent.

tanks on wheels (Fig. 8.15). This is typical for dairy farms. Under these conditions, the nutrient content is likely to be about one-third of that shown in Table 8.5.

The economic value of manure also is variable, depending on the nutrient percentages. In most cases, approximately one-half of the nutrients are released and available the first year. On this basis, if the analysis is known, a value can be placed on manure. Because manure has relatively low nutrient percentages, the volume that must be handled is relatively high if sufficient plant nutrients are to be applied. Consequently, manure is normally used fairly close to the farm or feedlot where it is produced. Manure application rates are generally 10 to 15 tons per acre (22 to 34 Mg/ha [megagram/hectare]). Manure should normally be applied on an "as is" basis for crops that have a significant nitrogen requirement.

Crops plowed under (Fig. 8.16) to improve fertility and physical condition of the soil are called green manure. The best crops for this purpose are legumes such as alfalfa and clover because they are high in nitrogen content; however, nonlegume crops such as wheat or sudangrass also can be used.

8.15.   A tractor-powered mobile tank and pump unit for injecting liquefied manure into the soil.

8.16.   Crops can be plowed under as green manure to provide organic matter.

Precision Farming

The computer age has led to many innovations in agriculture. One such innovation is precision farming. Because soils in most farm fields vary considerably in such properties as organic matter, topsoil thickness, texture, structure, and plant nutrient content, it is inefficient for an entire field to receive the same amount of fertilizer when the crop yield potential varies from one area of the field to another. It is claimed that fertilizer costs are reduced, the environment is better protected, and crop yields are higher if the proper amount of fertilizer is more precisely applied to each of the various soil types in the field. It can be demonstrated that the variations in soil properties are often more detailed than the soil map to be discussed in Chapter 12.

The *first step* in using precision farming techniques involves sampling the soil by using an ATV (all terrain vehicle) in a grid pattern with each sampling point commonly representing 2.5 acres (1 ha). (See Fig.8.5.) The position of each sample site is monitored and programmed into the computer via the GPS (global positioning system), which relies on a set of satellites about 200 miles (320 km) above the earth. The ATV is equipped with a receiver that picks up signals from at least three satellites for accurate positioning by triangulation (Fig. 8.17). With an enhanced system, the sampling sites can be pinpointed to within 1 to 3 feet (< 1 m). Each sampling location appears as a point on a computer screen.

The *second step* is to analyze the samples for nutrient content and other soil characteristics (such as pH and organic matter). From these results, a pattern of the fertilizer requirements throughout the field can be established and stored on a computer disk.

The *third step* requires a truck (fertilizer applicator) with three separate bins for nitrogen, phosphorus, and potassium fertilizers, each equipped with individual augers to transfer their contents to the spreading apparatus. The augers are regulated by a computer in the cab of the applicator whose position has been determined by the GPS. In this way, there can be a continuous adjustment in the rate and mixture of fertilizer applied as the truck goes back and forth across the field.

In the *fourth step*, the combine that harvests the crop is equipped with a device that continuously monitors the yield of the grain harvested from all parts of the field in relation to a previously programmed grid sequence described in the first step. The yield map is used to set the yield goal, which determines the amount of nutrient to apply.

Farmers may contract for the precision farming techniques described above, and many are currently doing so. The extent to which precision farming becomes a common practice depends on how farmers judge its economic value.

Organic Farming/gardening

Organic farming or gardening involves producing crops without applying commercial fertilizer or chemical pesticides.

8.17.   In precision agriculture, soil samples are taken from positions accurately located by satellite signals picked up by the receiver on the ATV.

Organic farming usually includes the slow release of a naturally balanced supply of nutrients from decaying organic matter (Fig. 8.18), such as crop residues and animal manure. Crops of high quality and quantity may be grown by organic methods. In parts of the world where commercial fertilizers are not available, farmers must rely on decaying organic matter for supplying crops with nutrients. Enough food could not be provided for the world's population if the plant nutrients for food crops were to be supplied solely from organic sources. The recycling of carbon and nutrients from plant and animal manures is, however, an important benefit of organic production techniques. Some consumers prefer to eat food grown without chemicals so they are willing to pay higher prices for crops that are grown using organic methods.

Nutrients from the soil are absorbed by plants mainly in the form of ions (weakly charged particles). Whether these ions come from a weathering mineral, decaying humus, or a chemical fertilizer is of no consequence to the plants. But the nutrients should be in a properly balanced proportion.

8.18. No inorganic fertilizers are used in organic gardening.

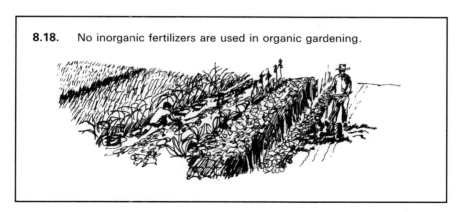

Composting

Composting is often used for a source of organic matter for gardeners (Fig. 8.19). The basic ingredients of a compost heap are organic residue, a little natural soil, moisture, nitrogen fertilizer, and some lime to counteract acidity associated with the nitrogen. The heap is allowed to rot for several months and is then mixed into the soil. The rotting process is carried out by microbial action that is greatly hastened in the presence of an adequate amount of nitrogen.

Composting is also becoming more popular for organic farming. A typical system has large rows of organic material (leaves, sawdust, or manure) that is amended with an organic source of nitrogen, and other nutrients if necessary. A machine mixes the material in the rows until the compost is stable (partially decomposed). The heating of the compost as some of the organic material decays has an added benefit of killing most of the pathogenic or disease-causing microbes. Finished compost can be bagged for sale in stores or loaded onto trucks or spreaders for application to fields.

Composting is a good way to turn waste products into a reliable and valuable plant nutrient source.

8.19. A compost heap disposes of organic waste and provides fertilizer.

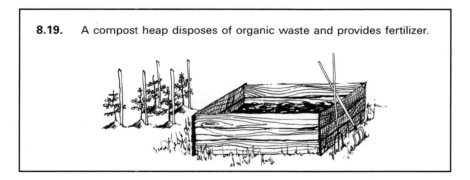

CHAPTER **9**

Soil Management

Proper soil management is an important part of any operation that involves crop production or natural resource management. The goal of soil management is to establish and maintain the correct combination of all soil factors necessary to optimize and maintain production efficiency. Effective soil management will ensure that food and fiber production are maximized and sustained over the years while leaving the soil in a productive state.

Soil management is closely related to crop management. It includes (1) maintaining the soil in a good physical condition; (2) keeping the chemical characteristics of the soil in the proper balance, such as maintaining soil fertility and the correct pH; and (3) influencing the biological or organic portion of the soil so that maximum benefits result.

Physical Condition

Soil with good physical condition is important to plant growth. The physical condition of soil as it relates to ease of tillage, quality of the seedbed, and resistance to seedling emergence or root growth is referred to as *soil tilth*. Good tilth helps the water infiltration rate, water-holding capacity, soil-air interchange, and root development, and it also aids in reducing erosion. Soil tilth can be maintained by continuing to return plant residues and organic materials to the soil and by using tillage practices that do not compact soil. Physical condition deteriorates if the soil is compacted by tilling when it is too wet or by using heavy machinery too often.

Residue Management

Crop residues influence the soil's physical condition. A soil that is loose and friable is generally considered to have good tilth. The first residue management step is deciding whether and when to incorporate crop residue. The time to incorporate residue will depend on the cropping system. Residue may be incorporated any time from immediately after harvest to just before planting of the next crop. If the field is to be planted the following year, it is sometimes desirable to incorporate residue soon after harvest to start the decomposition process. This helps to ensure that a smooth seedbed can be prepared and that the soil is free of excessive residue for planting. If a crop is not going to be planted the following spring, the residue can be left on the soil surface to act as a cover. This would greatly reduce erosion by both wind and water and conserve soil water.

Where wind or water erosion is a problem, partial incorporation of residue into the soil may be desirable, with a portion left on the surface (stubble mulching). Some farmers never incorporate plant residues into the soil and plant the subsequent crops with the residue still on the surface (minimum tillage).

As discussed in Chapter 4, bacteria decompose (break down or change) crop residue after it has been incorporated into the soil. Bacteria break down corn, wheat, and grain sorghum (all in the grass family) residues quite slowly, whereas alfalfa, soybean, and clover (all in the legume family) residues decompose more rapidly. The principal reason for the difference in decomposition between the two groups is the amount of nitrogen (in protein form) in the residue (Fig. 9.1). Legumes are high in nitrogen, whereas grasses are low in nitrogen. If it is ever desirable to speed up the decomposition or breakdown of a low-nitrogen residue, adding nitrogen fertilizer should help.

Reasons for Tillage

Tillage refers to the moving, turning, or stirring of the soil. The soil is tilled to accomplish a number of things:

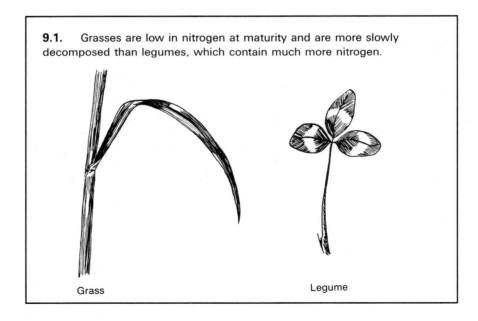

9.1.   Grasses are low in nitrogen at maturity and are more slowly decomposed than legumes, which contain much more nitrogen.

Grass   Legume

1. *Incorporate residue.* Incorporation of some crop residue into the soil hastens its decomposition. Without the decay of crop residues, it could become difficult to prepare a good seedbed and plant the seed. Different degrees of residue incorporation are illustrated in Figure 9.2; the implements employed are discussed in the next section.

2. *Improve physical condition.* Too often soils are worked or crops are harvested when the soil is too wet, causing the soil to compact or form a plow pan (Fig. 9.3). Some soils have naturally compact layers; deep plowing or chiseling can be done to break up the compacted soil or plow pan.

3. *Reduce erosion by wind or water.* Some tillage practices are used to reduce erosion (see Chapter 10). They include plowing on the contour, terracing, furrow diking, stubble mulching and sand fighting or creating ridges to roughen the soil surface.

4. *Prepare the soil for planting.* This may consist of cultivating beds in rows in which to plant seed. Or it may involve a light disking or harrowing to break up a surface crust and at the same time destroy small weeds.

5. *Incorporate pesticides, fertilizers, and animal manures.* Some pesticides and fertilizers may be left on the soil surface and still be effective. Most pesticides, however, are incorporated either with a light disking or by using a rotary hoe. Fertilizers and manures are often incorporated into the soil by either disking or plowing.

6. *Control pests, including weeds, insects, and diseases.* Even though many types of weeds are controlled by chemicals, tillage practices also are used. A light disking before (or during) planting may be used to kill early-emerging weeds. Deep moldboard plowing is occasionally used to turn up roots of hard-to-kill weeds such as johnsongrass so that they will be killed by freezing. Shallow tillage may be used on fallow land to control weeds. Tillage also is used to incorporate residues or host plants that might harbor insects and diseases.

7. *Increase water infiltration.* In areas where moisture is often limiting or where irrigation is practiced, certain tillage methods are used so that water, particularly rainfall, can move into the soil more rapidly or stay on the surface until it does so. Tillage that breaks up a surface crust can increase the infiltration rate. Forming rows on the contour and terracing as well as furrow diking (see Chapter 10) helps soil hold water longer.

Tillage Implements

Several types of farm implements are used to accomplish the tasks listed above. The degree of soil disturbance varies by the type of implement, how deep it is placed in the soil, and how fast it passes through the soil. *Primary tillage implements* refer to those that disturb the soil a great deal and incorporate significant amounts of crop residue (moldboard plow, chisel plow, and disk), while *secondary tillage implements* (disk, field cultivator, and harrow) generally till the soil to a shallower depth and often follow primary tillage operations.

9.2. Disking incorporates crop residue to a shallow depth, a moldboard plow covers the residue, and a chisel plow goes deep but leaves no residue on the surface.

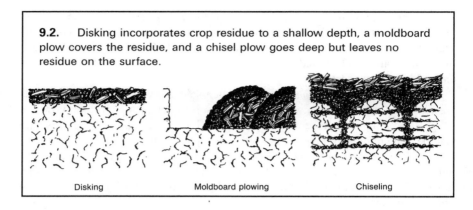

Disking   Moldboard plowing   Chiseling

9.3. Plow pans can form at the depth of tillage and inhibit root penetration because of their increased density. Chiseling or periodic deep plowing can prevent this effect.

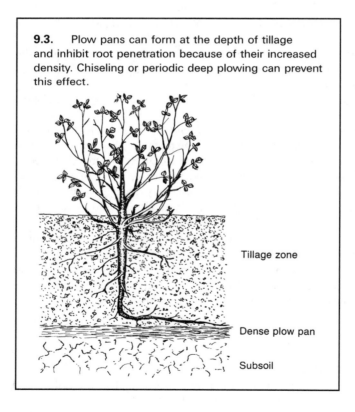

Tillage zone

Dense plow pan

Subsoil

A ***moldboard plow*** (Fig. 9.4a) is used to lift the soil and completely or partially turn it over. This can be done at any depth, but in most areas, it is 6 to 10 inches (15 to 25 cm). Development of a moldboard plow with steel shares was one of the key factors that enabled development of the Midwest

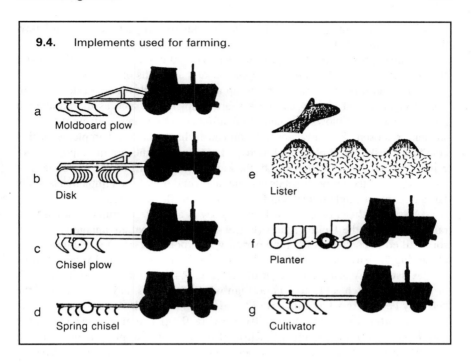

9.4.   Implements used for farming.

a   Moldboard plow

b   Disk

c   Chisel plow

d   Spring chisel

e   Lister

f   Planter

g   Cultivator

and Great Plains of the United States as major crop-producing areas in the 19th century.

Moldboard plowing was often done to incorporate residue and break up dense soil. When followed by some secondary tillage, moldboard plowing helped create a fine seedbed for planting. Moldboard plowing incorporates most of the crop residues, which may be good for residue decomposition and seedbed preparation but can also leave the soil vulnerable to erosion. It was once the most common type of primary tillage but its use has decreased significantly due to the high cost of fuel to pull the plow and concern regarding excessive erosion.

Deep plowing to a depth of 18 to 36 inches (46 to 92 cm) is done in certain areas. This is usually used on a soil that has a sandy surface and a clay subsoil, with clay being brought to the surface to mix with the sandy portion. In some regions, it also is used to incorporate underlying mineral soil into a surface layer of thick muck (high organic material).

A **disk** (Fig. 9.4b) can be used for both primary and secondary tillage by varying the depth of the implement. It is often used to incorporate materials into the surface of the soil, usually the top 4 to 6 inches (10 to 15 cm). This often includes chemicals for weed control, fertilizers or other soil amendments, or crop residues. A light disking also is used to control weeds. Disking or shredding of crop residues with an implement called a *flail chopper* sometimes precedes moldboard plowing to improve residue incorporation and prevent clogging of the plow. One disadvantage of disking is that it tends to compact the soil just below the depth being tilled.

A **chisel plow** is another type of implement used in primary tillage (Fig. 9.4c). It is an implement pulled through the soil usually at depths of 10 to 14 inches (25 to 35 cm) and used primarily to break up a hardpan or plow sole (a dense, compacted layer of soil usually caused by farm implements such as the moldboard plow). Some chisel plows are made so that the chisel vibrates, causing the soil to shatter or loosen. Recently, several companies have produced implements that combine a disk, chisel plow, and harrow. These implements are quite popular because the depth, spacing, and size of the individual elements can be customized or adjusted to suit the farmer's needs and the field conditions. Such implements allow farmers to use a single implement for most of their tillage operations.

A **spring chisel** or Graham-Hoeme plow is a tillage instrument in the low-rainfall areas of the United States (Fig. 9.4d). Its principal advantage is that the soil is tilled to control weeds (and partially incorporate residue) while minimally disturbing the surface soil and thereby reducing loss of soil water by evaporation.

A **lister** may also be used for tillage (Fig. 9.4e). It forms the soil into beds or rows 6 to 8 inches (15 to 20 cm) high where the seeds may be planted or the bed may be lowered 3 to 4 inches (7.5 to 10 cm) before planting.

A **planter** (Fig. 9.4f) and a **cultivator** (Fig. 9.4g) also may be used to till the soil. Even though the principal function of the planter is to place the seed in the soil, the furrow openers and other attachments are capable of cutting through crop residues so fertilizers and pesticides can be applied during planting. Cultivating was usually done in the past to control weeds that germinated after planting. However, modern herbicides have replaced much of the requirement for cultivation. Culivation is still used in some areas when chemical weed control fails and to break up crusts to improve infiltration.

Tillage Practices

Tillage practices used to manage soil vary widely. The type of crop grown, type of soil, erosion hazards, the use (or not) of irrigation, and cost of the tillage practice are all considered in determining which practices to use.

Tillage of soil may vary from farms on which there is no tillage (except planting)—called *no-till or minimum till* systems—to farms where the soil may be tilled 8 to 10 times a year. Because of excessive erosion and high costs, the trend for most producers is to reduce the number of tillage operations. Many farmers now use a *reduced or conservation tillage* system, where some crop residue is left on the surface and the soil is only moderately disturbed. These systems can be thought of as a compromise between the extremes of tillage, moldboard plowing and no-till. Reduced tillage was initially adopted mostly by farmers in the higher rainfall areas of the eastern and midwestern parts of the United States.

In a no-till system, all of the crop residue is left on the surface and seeds are planted with as little disturbance of the residue and soil as possible. An example of no-till planting is illustrated in Figure 9.5. In addition to cost savings, minimum tillage also decreases soil erosion and normally increases the water infiltration rate. Minimum tillage may require a modification of certain practices, including planting, fertilization, and application of pesticides. Fertilization rates, particularly for nitrogen, may need to be increased because soil temperatures tend to be lower as a result of the surface residues. Surface residue also could harbor insects, thereby resulting in an increased need for insect control.

Soil erosion by water and wind has been a problem since soils have been tilled. Conservation tillage practices were developed to reduce the loss of valuable topsoil by erosion; thus, they often involve leaving a portion of crop residues on the surface of the soil. By definition, conservation tillage is in effect when 30% of the soil surface is covered with crop residue. Conservation tillage has been practiced for many years in the areas of limited rainfall and potential wind erosion in the western United States. Crop residue is left on the surface of the soil not only to decrease erosion but also to conserve water (stubble mulching).

Chemical Characteristics

It is also desirable to keep the chemical characteristics of the soil in proper balance or condition. This includes maintaining soil pH in an optimal range, providing a sufficient and balanced supply of nutrients, preventing or alleviating saline-sodic soil conditions, and avoiding soil degradation from toxic pollutants.

Soil pH

Soil pH (see Chapter 5) is important in crop production and is an indicator of the acidity or alkalinity of a soil as well as an indicator of levels of certain nutrients and their availability. It also influences biological activities in the soil.

9.5. Minimum tillage or no-till often means planting while residue from the previous crop is still in the field.

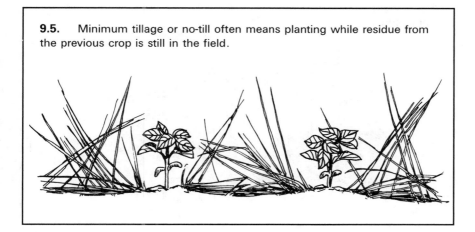

If a soil is too acid (below pH 5.0), phosphorus, iron, and certain other nutrients have limited availability, and levels of calcium, magnesium, and potassium would normally be low. If a soil is too alkaline (above pH 7.8), phosphorus, iron, zinc, manganese, and other nutrients might have reduced availability. On the other hand, calcium, magnesium, and potassium would normally be high.

Soils in the eastern United States and in many of the high-rainfall areas worldwide are more likely to be acid. If soils in these areas are in crop production, the pH usually ranges from 7.0 to around 5.0. A soil test is the best method for determining whether pH needs to be adjusted. Most soils are not adversely affected by acidity if they are in the pH range of 6.0 to 7.0. If the soil pH is 6.0 or below, it may be necessary to raise the pH (make it less acid) by the use of limestone (calcium and/or magnesium carbonate). Limestone for application to soil (also called aglime) is relatively inexpensive because it can be quarried from abundant deposits (Fig. 9.6).

The amount of limestone needed per acre (hectare) depends on the cation exchange capacity (CEC) of the soil (see Chapter 5). The CEC is dependent on texture, type of clay mineral, and organic matter content. At a pH of 5.5, a soil with a high CEC might need 3 tons of limestone per acre (6.72 Mg/ha [megagrams/hectare]) every 2 to 3 years, whereas a soil with a low CEC may need only 1 ton per acre (2.24 Mg/ha) to raise the pH to a desired level (Fig. 9.7). The application rates and amount of time between applications depend on the crops being grown, the amount of rainfall and leaching, and the CEC.

Limestone increases pH (decreases the acidity) by providing calcium, and in some cases magnesium, which in turn replaces hydrogen or other acidic ions on the exchange complex. With more calcium and magnesium, which are bases, and less hydrogen on the exchange complex, pH increases.

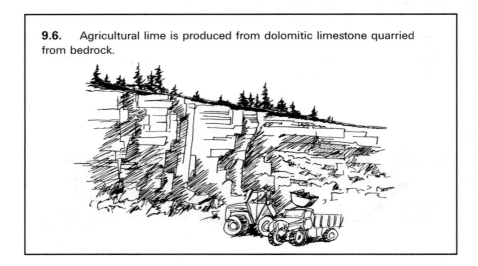

9.6. Agricultural lime is produced from dolomitic limestone quarried from bedrock.

9.7. Many humid-region soils need regular applications of lime to combat acidity.

In the western United States and other limited rainfall areas of the world, soils tend to be neutral to alkaline. If the soil pH needs to be decreased (made less alkaline) and the soil is not sodic (to be discussed later), elemental sulfur (S) may be applied. Using sulfur to decrease the pH of the entire soil mass would be quite costly, hence, the economic benefits usually would be less than the returns. Where high pH due to calcium is a problem, the common practice is to apply sulfur or an acid-forming sulfur product into a small band or limited soil area. The sulfur forms an acidic microenvironment in which nutrient availability may be greatly increased. In this way, a small amount of sulfur at a low cost can be beneficial for one season in counteracting the undesirable effects of high pH. The amount of sulfur required to effect this change in a microenvironment may range from 20 to 100 pounds of sulfur per acre (22 to 110 kg/ha) if properly applied. Sulfur products most commonly used for this purpose are prilled sulfur (80% to 90% elemental S), ammonium thiosulfate (26% S), and ammonium polysulfide (45% S).

Nutrient Supply

Keeping plant nutrients at adequate levels is important and proper nutrient balances need to be maintained to ensure that nutrient levels are adequate but not too high. Next to irrigation, this is probably the single most important soil management factor that can influence yield over which the producer has control. If too much of a nutrient is applied, not only is it an inefficient use of resources (Fig. 9.8) but it also may alter the balance of nutrients, adversely affect plant growth, and create a potential source of pollution. For a thorough discussion of nutrients and their application, see Chapter 8.

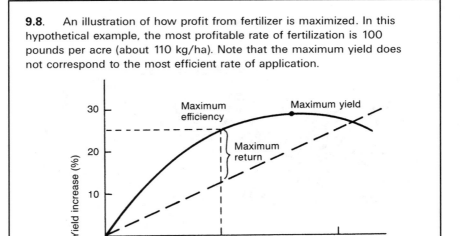

9.8.   An illustration of how profit from fertilizer is maximized. In this hypothetical example, the most profitable rate of fertilization is 100 pounds per acre (about 110 kg/ha). Note that the maximum yield does not correspond to the most efficient rate of application.

Saline and Sodic Soils

Another important management practice is to reduce the effect of saline and sodic soils on plant growth. A *saline soil* is one in which soluble salts have accumulated in sufficient quantity to adversely affect growth. A *sodic soil* is one that contains too much sodium, which adversely affects yields. Remediation of saline and sodic soils may involve a combination of soil management practices.

Saline and sodic conditions may occur naturally, but most arise when irrigation water is applied that is too high in salt and/or sodium. Areas in the United States where these situations occur are mainly in the Southwest, from Texas to California. It is a potential hazard in any irrigated area in the world.

Irrigation waters can be chemically analyzed to determine whether salt and/or sodium are high enough to create problems if used. Chemical tests are highly desirable for any new irrigation project or in areas where the salinity of water might tend to change.

Soils that become saline show irregular growth of crops in a field and usually have a whitish cast from salt accumulation, with the greatest amount being in the tops of the beds (Fig. 9.9A). Saline soils are often called "white alkali" because they are light in color and have an alkaline soil reaction.

If saline conditions develop, leaching the soil with water with less salt is a common practice. The downward movement of water in the soil that occurs during leaching carries the salts below the root zone, where they cause less of a problem.

If a soil becomes sodic, it is highly dispersed and in very poor physical condition (see Soil Aggregation in Chapter 5). The soil feels slick and gummy when wet. When dry, the soil is dark, appears to be highly dispersed, and has cracks.

9.9. Saline soils (A) usually have "white caps" of salt in the tops of the beds. Growth of crops normally is spotty. Sodic soils (B) are usually dark colored (often called "black alkali") and are gummy and slick when wet and cracked with a powdery surface when dry.

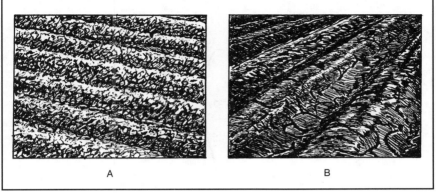

A   B

Sodic soils are called "black alkali" because of their dark color and alkaline soil reaction (Fig. 9.9B). Sodic soils usually range in pH from 8.5 to 10.0.

To correct a sodic soil condition, a calcium-containing compound—specifically gypsum, which is calcium sulfate ($CaSO_4 \cdot 2H_2O$)—would need to be applied. The calcium would replace the sodium by combining with the sulfate. Leaching with high-quality water is required to move the sodium sulfate downward and out of the root zone. The calcium then helps the soil to reaggregate and improve in physical condition. In some sodic soils, calcium may be present as calcium carbonate and, if so, only sulfur (an acid-forming type) is needed to change the sodic situation. The general pH management considerations for the soils of the United States are summarized in Figure 9.10.

Biological Characteristics

Life in the soil, or soil biology, was discussed in Chapter 4. One of the principal practices related to the biology of the soil is crop residue management. Crop residue decomposes to replenish soil organic matter, a process that has important and generally beneficial effects on soil biological activity. Other management practices affecting soil biology include the application of animal manure or other organic materials such as wastewater treatment plant biosolids and cannery or processing plant byproducts. Physical or chemical conditions in the soil that adversely affect soil organisms (poor aeration or toxic levels of chemicals) should also be avoided.

Plant residue also serves as a source of energy for organisms that live in the soil. As crop residue decomposes, nutrients that plants require are released.

9.10.   The generalized pH management considerations for soils of the United States. Region A has many areas of saline and sodic soil conditions. In region B the acid-base relationships are commonly favorable, and in region C the bases have been leached so that lime and fertilizer are needed in high amounts.

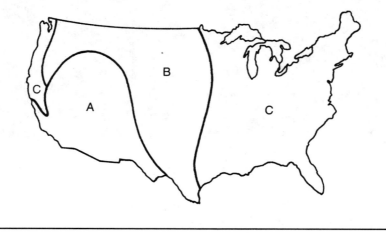

Organic acids are formed that in turn may enhance the availability of certain plant nutrients and also aid in the breakdown of minerals in the soil.

    After humus is formed, additional cation exchange capacity is present. Organic matter helps a soil retain more water. Water infiltration rate often increases with additional organic matter as a result of an improved aggregation of soil and enhanced structure. Soil temperature also may tend to be slightly higher because increased organic matter generally causes a soil to darken. Other benefits of increased organic matter include (1) reduced toxicity of certain pesticides, (2) increased buffering in the soil, and (3) a decreased effect of saline-sodic soil conditions.

    Crop residue on the surface decreases the detrimental impact of raindrops and thereby decreases erosion caused by runoff water. Residue cover also decreases erosion by wind by slowing the wind near the soil surface. In winter, crop residue can trap and hold snow. As the snow melts, the water moves into the soil where it can be stored and taken up by the crop later. In areas of limited rainfall, this extra moisture may be critical for crop production. Surface residue also conserves moisture by keeping soil temperatures lower and decreasing the loss of water by evaporation.

    Crop residue can have some potentially detrimental effects. It may harbor diseases that would appear the following year. Weed seeds also may be present and thus may germinate the next year. If the residue is low in nitrogen, which is normally the case with crops like corn and wheat, temporary nitrogen shortages may occur for the next crop to be grown in that soil. Residue is often burned to avoid one or more of the above problems. Normally, it is undesirable to burn residue; however, the economics of a situation may dictate that it be destroyed. Burning residue year after year is definitely disadvantageous, but once every 5 years or so will not appreciably lower soil organic matter content. Burning of residue is a common practice with rice (Fig. 9.11). In California, some rice straw is baled and injected with ammonia for animal feed. This helps to reduce air pollution from smoke, which has become an environmental issue.

Crop Production Factors

Crop production is influenced by many factors; some can be controlled by the farmer and some cannot. These factors can be classified as (1) soil, (2) crop, (3) environmental practices, and (4) cultural practices.

The many facets of the soil are discussed throughout this book. Its physical, chemical, and biological characteristics should be maintained under optimum conditions for best crop growth.

The crop to be grown involves many management decisions. A crop needs to be adapted to the soil as well as to climatic conditions. Markets also must be considered. Choice of cropping system (whether to grow one crop continuously or to use a particular cropping sequence) is important. A cropping system may involve continuous corn in the midwestern United States or a sequence of corn/soybeans every 2 years. In the Southwest, a cropping system may be continuous cotton or a sequence of cotton, grain sorghum, and wheat over a 4-year period.

Other management decisions on crops include which variety to plant. Some varieties are adapted for high-yielding potentials where fertility and moisture are

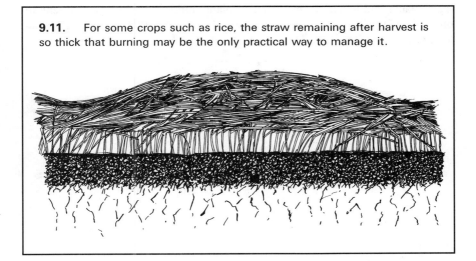

9.11. For some crops such as rice, the straw remaining after harvest is so thick that burning may be the only practical way to manage it.

adequate, whereas others are better adapted to low potential yield situations such as in dryland production areas. The density of certain crops also may be important and needs to be maintained at an optimum level.

The environment is generally the most important aspect of crop production and often cannot be controlled or influenced. Length of growing season, altitude, day length, light energy, and rainfall are among conditions over which the farmer has no control. Even though rainfall cannot be changed, certain management practices can be used to conserve water. Contour rows, crop residue cover, furrow diking, and similar practices can be used to increase retention of the water that does fall.

For some specialty crops, management practices can be used to modify the plant growth environment. For example, additional lighting can be used to control the growing season, day length, and light energy for greenhouse plants, but little can be done for most field-crop production. An exception would be tobacco grown under cheesecloth, where the amount of light reaching the plant is reduced.

Using the right cultural practices at the right time (management) is important. Such decisions start after harvest and could include (1) whether to incorporate residue or leave it on the soil surface; (2) what tillage practices to use; (3) when to plant and how deep to plant; (4) the type and rate of fertilizer and how and when to apply; (5) the rate, time, and method of application of herbicides, insecticides, or fungicides for pest control; (6) when to apply irrigation water if it is available; (7) when to harvest; and (8) what marketing strategy to use. And the alternatives could go on and on. A producer makes many management decisions each year, with almost every decision being a critical one for yield and profit.

Conditions that could limit yield are numerous. A few are cited here as a checklist to determine why yields are less than the potential: (1) biological hazards such as weeds, diseases, or insects; (2) nonbiological hazards such as hail, excessive or insufficient rainfall, early or late frosts, or extreme temperatures; (3) inadequate stand level; (4) imbalance of plant nutrients or improper nutrient application; (5) poor physical condition of soil; (6) soil pH that is too high or too low; or (7) improper variety (cultivar).

CHAPTER 10

Soil Conservation

If rains are gentle on native grasslands and forests, the water soaks into the soil and percolates downward through the soil profile. If there is enough water, it eventually penetrates to the water table. When rain falls faster than the soil can absorb it, some water soaks in and the rest runs over the soil surfaces to the lowest part of the field and eventually runs off. The runoff water from vegetated land may look clear, but it is actually carrying some mineral and organic matter derived from exposed soil, earthworm casts, and the digging activities of ants, moles, badgers, and similar organisms. This process represents normal or geologic erosion. In some places, geologic erosion may lower the ground surface only an inch (2.54 cm) in 5,000 years. This is probably a normal rate for hills and valleys to form. In desert areas where vegetation is sparse, the rate may be no faster because rain is infrequent and the soil surface is typically armored with a layer of pebbles. Landslides have always taken place in mountainous regions. Geologic erosion also includes the collapse of riverbanks into floodwaters during periods of excessively high rainfall.

European immigrants to the United States cleared the native vegetation to make cultivated fields. Their livestock often overgrazed rangelands, which weakened the plant cover and trampled the soil. As a result, the original condition of the soil deteriorated rapidly and erosion intensified.

The word *erosion* literally means gnawing away. Soil erosion, as opposed to geologic erosion, is the accelerated washing and blowing of soil as a result of its disturbance by humans.

In regions where the soil parent material is loose, such as where there is deep glacial till, the long-term effects of erosion are serious but not as devastating as where solid bedrock is the parent material (Fig. 10.1). The latter includes most of the earth except for the valleys, which have been filling with sediment over millions of years.

Soil erosion is a serious problem on much of the world's cultivated land (Fig. 10.2). A look at most large rivers shows us they carry a heavy load of sediment that comes from watersheds. A ***watershed*** is all the land area that yields water from rain and snow to a particular river. A watershed need not become a "soilshed" too. In arid regions where vegetation is sparse and rains are quite often intense, erosion may be severe without human involvement. But in areas where a protective vegetative cover is normal, it is usually human activities and mismanagement that have accelerated erosion. This most often involves the removal of vegetation by tillage or overgrazing by livestock. Most agricultural systems tend to accelerate erosion, but it is the obligation of those who manage farms, ranches, and forests to keep erosion to a minimum.

10.1.   Loose substratum (A) slowly develops into soil if surface erosion takes place at a slow rate. Where soil is thin over bedrock (B), erosion of the surface leaves a barren landscape.

10.2.   Agricultural systems commonly accelerate erosion.

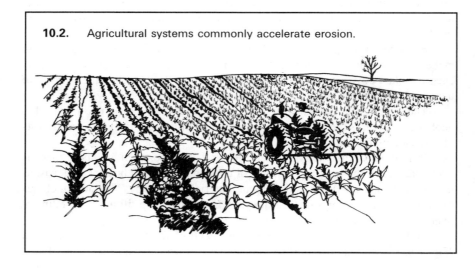

Erosion by Water

Both water and wind erosion consist of the processes of detachment, transportation, and deposition. Soil erosion by water occurs when particles are loosened (detached) and carried (transported) by moving water. The soil particles are eventually deposited. As long as soil particles are clustered into aggregates, they are not easily moved. But if the soil is exposed to the bombarding force of raindrops, the aggregates tend to break apart, and the detached particles are subject to movement in the runoff water (Fig. 10.3).

It has been calculated that the rate of fall of a raindrop is about 20 miles (32 km) per hour. The kinetic energy generated by a 2-inch (5-cm) rain on 1 acre (0.46 ha) is about 6 million foot-pounds (4.4 million joules). This is enough energy to raise a 7-inch (18-cm) layer of soil 3 feet (0.9 m). Clearly, this is not all used to move soil, but it does show that the impact of raindrops releases a large amount of energy that contributes to soil detachment and erosion. On a 10% slope, 60% of the soil splashed by raindrop impact moves downslope and only 40% is thrown upslope. The net movement downhill is called *splash erosion*.

Types of Water Erosion

The three main types of water erosion are gully erosion (gullying), rill erosion, and sheet erosion.

The most spectacular type of water erosion is **gullying**, which occurs when water concentrates in a channel and deepens it rapidly (Fig. 10.4). This is what happened over a long period to the Grand Canyon of the Colorado River, and it is happening on a smaller scale on thousands of farms throughout the world. A gully generally starts at the outlet of a channel and works its way upstream by waterfall action at its head. The gully extends itself upstream by undercutting the floor of the channel. On a larger scale, Niagara Falls also is working its way upstream, as shown by the occasional fall of masses of limestone bedrock from the rim.

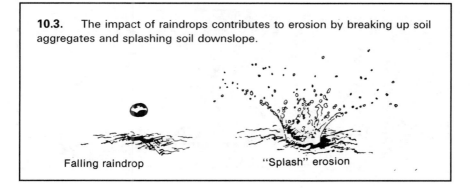

10.3. The impact of raindrops contributes to erosion by breaking up soil aggregates and splashing soil downslope.

Falling raindrop   "Splash" erosion

10.4.   Gully erosion can be spectacular.

Rill erosion consists of the removal of soil on a sideslope by small channels (Fig. 10.5) that are not deep enough to interfere with tillage equipment. **Sheet erosion** is the planing off of a land surface by water action without formation of channels. This generally happens where there is not enough cover of vegetation over the soil to prevent erosion but enough to prevent rilling.

Rill and sheet erosion go unnoticed because tillage destroys the evidence until most of the topsoil is gone and the subsoil is exposed at the surface. Although less spectacular than gully erosion, rill and sheet erosion cause the loss of a great deal more soil.

10.5.   Rill and sheet erosion can result in great soil loss.

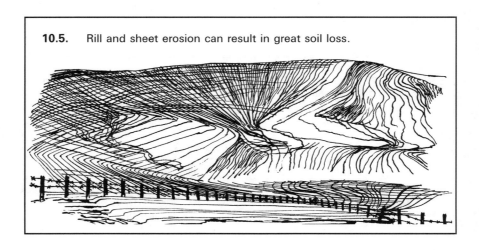

Water Erosion Control

It is an old adage that "an ounce of prevention is worth a pound of cure." This certainly applies to soil erosion control. Because soil particles do not move until they are detached, every effort should be made to prevent this from happening in the first place. Two factors should be kept in mind in this regard. The first is that the strong, water-stable aggregates usually associated with high organic matter content allow water to infiltrate harmlessly into the soil. Second, a cover of vegetation or plant residues functions to dissipate the energy of raindrops so they cannot strike directly on the soil aggregates (Fig. 10.6).

The fundamental principle of preventing the transportation of detached soil particles by runoff water is to reduce the rate of flow down the slopes. Slow-moving water does not have the energy to transport a large load, and the slower speed allows more time for the water to infiltrate.

One of the common ways of controlling runoff is to subdivide the sloping fields into contour strips (Fig. 10.7). Each strip is approximately perpendicular to

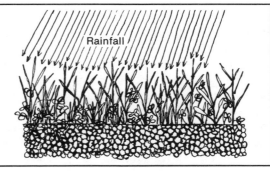
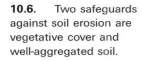

10.6. Two safeguards against soil erosion are vegetative cover and well-aggregated soil.

Rainfall

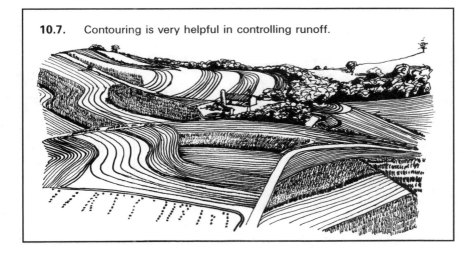

10.7. Contouring is very helpful in controlling runoff.

the downhill path the water would take. Alternate strips usually contain a row crop such as corn in one strip and a forage or small grain crop in the next row. The ridges in the strip crop reduce the water flow and minimize erosion there, and little erosion can occur in the dense plant cover in the alternate strips. No-till planting, in which the previous year's stubble is left on the surface, is another practice that reduces erosion even when it is not done on the contour. It permits use of agricultural equipment that is too large for contour strip-cropping operations.

Where water concentrates in channels and flows in an uncontrolled manner, the channels may become gullies. Grassed waterways (Fig. 10.8) can reduce the risk of gullying because the dense fibrous root system of grasses holds the soil securely against the force of the flowing water. Grassed waterways may be planted in natural channels that lead downslope, in diversion ditches on the sides of slopes, and at the foot of adjoining slopes.

If the natural slope of the channel is too steep, it must be flattened out between steel or concrete structures called drop spillways (Fig. 10.9), which act as steps in the channel. At each step, the water drops over a headwall and falls on a concrete apron from which it flows onward through a grassed waterway.

In commercial agriculture today, conservation practices have advanced to the point that gullies are seldom a major problem. The size of farm machinery is such that if gullies start to form, they are easily removed. If gullies continue to form, however, farmers often are forced to divide their fields inconveniently because gullies cannot be crossed with machinery. Fences are undercut, and roads have to be rerouted around advancing gullies. In some instances buildings have toppled into them.

To stop the advance of a gully, the soil must be stabilized at its head where the waterfall action is occurring. Sometimes the head of the gully is filled in with concrete or rubble (riprap), and the sides and bottom are planted with trees, grass,

10.8. A grassed waterway offers erosion protection.

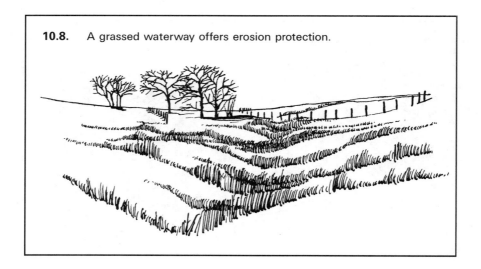

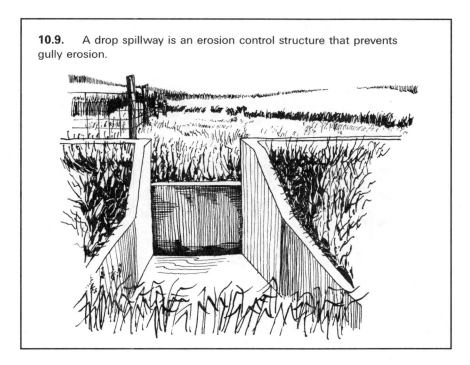

10.9. A drop spillway is an erosion control structure that prevents gully erosion.

and shrubs. These practices may be combined with the use of a chute through which runoff water passes to prevent any further undercutting at the gully head. Various other techniques may be employed, depending on the specific problems at hand.

A principal point to keep in mind is that in many cases it is desirable to keep as much of the water as possible where it falls or is applied by irrigation so that it can move into the soil and be used by plants. This is very important in arid and semiarid areas where moisture is often the limiting factor. Contouring and strip cropping are techniques in this category. Another approach is furrow diking, which is an old practice that is now being used much more commonly in dryland areas (Fig. 10.10). A dike is placed in each furrow at intervals of 6 to 12 feet (1.8 to 3.6 m), with a resulting depression holding water between the dikes. This greatly decreases runoff and improves plant growth. In the furrows the tractor follows during harvest, the dikes must be removed by sweeps mounted in front of the tractor.

Another mechanical erosion control practice is terracing. There may be no older erosion control method than the construction of terraces. In some parts of the world, entire mountainsides have been modified into a series of steps by bench terraces, each one supported by a retaining wall made of stones. Tremendous amounts of labor were required to construct them, and they still require labor to maintain, which is very cheap in some areas. Terraces on a much more

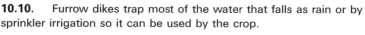

10.10.   Furrow dikes trap most of the water that falls as rain or by sprinkler irrigation so it can be used by the crop.

modest scale have been used compatibly with modern agriculture. In these cases, ridges are built several yards (meters) apart across the slope of the hills to trap water falling on the soil between the terraces. Presently, the trapped water is usually allowed to flow down the hill through buried tiles so that erosion does not occur (Fig. 10.11). In drier areas, the water may simply be held above the terrace until it soaks into the ground.

Erosion by Wind

Whereas the loss of soil by water erosion is more widespread than by wind, it was the devastating wind erosion of the drought years of the mid-1930s in the United States that focused attention on all types of soil erosion and brought about legislation dedicated to soil conservation (Fig. 10.12).

In the early part of the 19th century, much of the prairie in the Midwest was plowed with huge steam tractors and planted to wheat. In addition, much of the

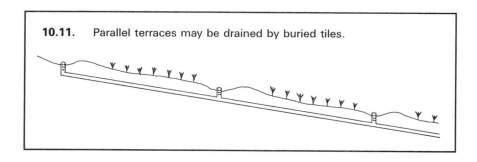

10.11.   Parallel terraces may be drained by buried tiles.

10.12. Although less common than water erosion, wind erosion can be devastating.

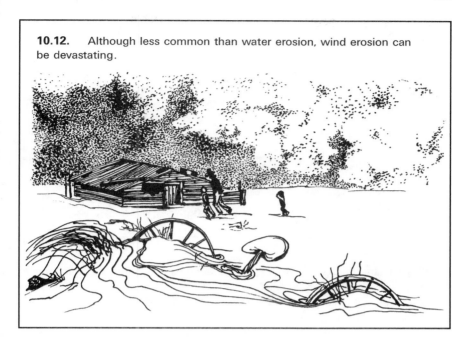

unplowed land was overgrazed so that the soil also was left with an inadequate vegetative cover to afford protection. When the drought and winds came in the 1930s, soil from the Great Plains was blown aloft in ominous dark clouds that sometimes reached the Atlantic Coast and even out to sea. Fences were covered, road ditches were filled, and farmyards were smothered with eroded soil, and there was much despair in the hearts of farmers and city folk alike. Even today, wind erosion continues to move vast quantities of the most fertile topsoil, thereby reducing the productive capacity of cropland.

Types of Wind Erosion

Wind erosion occurs when the wind is sufficiently (a) strong to move soil particles along the soil surface; (b) turbulent to keep particles suspended; and (c) gusty to keep soil moving. There are three types of wind erosion: saltation, surface creep, and suspension (Fig. 10.13).

Saltation is the bouncing of medium and fine sand particles along the surface after they initially start to roll. **Surface creep** is the rolling of coarse sand particles on the soil surface. Damage occurs to growing plants when surface creep and saltation occur. Sand particles constantly bombard the growing plants and eventually kill them if the wind is severe.

The abrasive action of the sand particles loosens finer soil particles, silt and clay, and brings about the third type of wind erosion, **suspension**. These fine soil particles, silt and clay, are often suspended in the air, resulting in a dust storm,

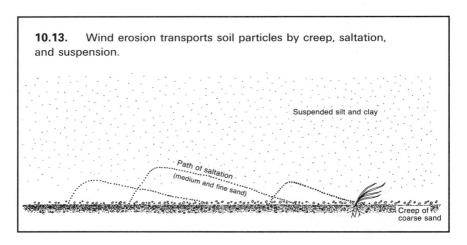

10.13.   Wind erosion transports soil particles by creep, saltation, and suspension.

and may be carried hundreds of miles. Although this may not be too damaging to growing crops, the loss of valuable topsoil with its organic matter and nutrients is costly. Damage also occurs when the soil particles are deposited. Deposition may be at the edge of a field where sand is deposited or many miles away from the origin of the particles. Recent studies have shown that dust from deserts in Africa blows across the south Atlantic and supplies nutrients for plant growth on otherwise infertile soil in the Amazon basin of South America.

Control of Wind Erosion

Control of erosion by wind is achieved by slowing down the wind, much the same as controlling erosion by water is accomplished by slowing the runoff. Vegetative cover is the most effective means of control—simply keep the soil covered so that it cannot be blown away. The vegetation may be alive in the form of a growing crop cover, or it may be dead plant residue sheltering the soil (Fig. 10.14). Often crop residue of this nature is partially incorporated but a portion is left on the surface to prevent blowing of the soil. The common method of tillage to achieve this stubble mulching is shown in Figure 10.15.

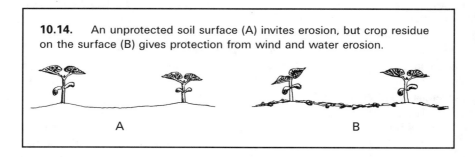

10.14.   An unprotected soil surface (A) invites erosion, but crop residue on the surface (B) gives protection from wind and water erosion.

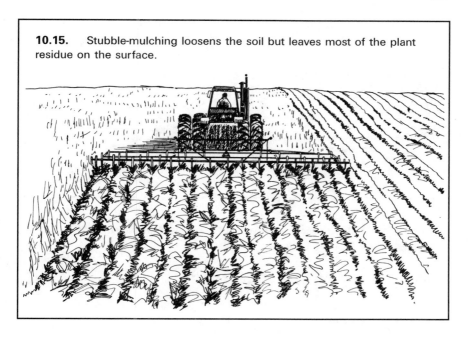

10.15. Stubble-mulching loosens the soil but leaves most of the plant residue on the surface.

Leaving the soil in a roughened condition also helps control wind erosion. Any obstacle in the way of a moving soil particle deters it and, in many cases, stops it. The obstacle might be cloddy soil, pitted areas in the soil surface, or ridged rows that have been prepared. Farmers in the Great Plains of the United States use a sand fighter, which is an implement taken over a field after a rain when the soil surface is smooth to create thousands of small pits and mounds to impede the movement of soil. This is usually done in the spring when wind speeds are highest and crops have just been or are soon to be planted. Using a sand fighter may kill a few plants if the crop is up, but damage is minimal.

Strip cropping is another means of minimizing soil movement by wind. The pattern would be similar to that used for erosion control by water, such as preparing a seedbed in the standard method for 8 to 12 rows but leaving 1 to 2 rows of the old crop standing. Strip cropping is often used on vegetables grown following a volunteer wheat crop.

Windbreaks, or shelterbelts (Fig. 10.16), are used in many areas where erosion by wind is a problem. A planting of trees, shrubs, or grass strips helps to slow the wind and decrease soil movement. To be effective, the distance between windbreaks should not be too great. The distance for best control varies from one area to another. Contact your local Natural Resources Conservation Service personnel for specifics.

Soil stabilizers also might reduce erosion by wind. These include chemicals that cause soil particles to aggregate, organic matter that acts as a binding agent,

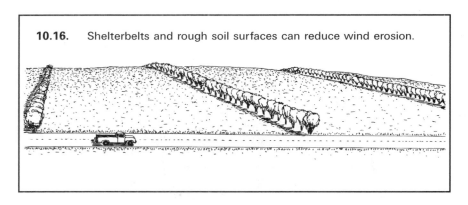

10.16.   Shelterbelts and rough soil surfaces can reduce wind erosion.

and water. Water stabilizes the soil for a time, but when the surface dries out, it is again susceptible to wind erosion.

Erosion by Mass Wasting

Mass wasting is another type of erosion that takes place in several ways. Masses of soil sometimes move under the force of gravity, as by cave-ins along riverbanks (Fig. 10.17) and slump along gully sides and roadbanks. Soil creep is a common type of mass wasting in regions where the soil freezes and thaws repeatedly each autumn and early spring (Fig. 10.18). Frost causes the surface soil to expand perpendicular to the slope; when the soil thaws, it drops vertically. This causes a ripple effect on the hillsides. Cattle often make paths that accentuate these ripples. Mass wasting not only may remove portions of cultivated fields and pastures but also sections of roads and even homes (as in the mudslides in California).

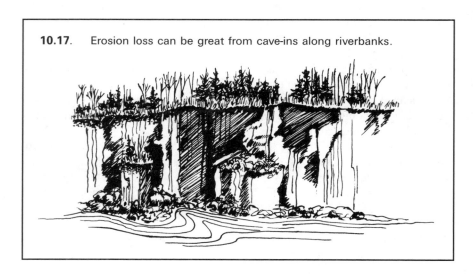

10.17.   Erosion loss can be great from cave-ins along riverbanks.

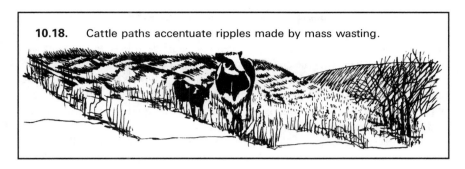

10.18. Cattle paths accentuate ripples made by mass wasting.

Diversion of water away from susceptible areas is helpful. Protection of riverbanks with riprap, timberwork, and vegetation, along with careful channelization, reduces the incidence of mass wasting. Similar protection also is being sought for shorelines and beaches; huge sand-filled plastic tubes are laid at the foot of an endangered bluff at a lake or seashore to block the force of waves.

Sediment as a Pollutant

Whenever soil is moved by wind, water, or gravity, it is likely to cause immediate problems in addition to those of a more long-range nature. The occurrence of massive mudslides in such places as California usually attracts widespread media attention. This accelerated erosion is often attributable to the loss of vegetation due to fire or logging. Houses may be filled with the flowing ooze or even swept down the hillsides.

Sand blown along by the wind is a serious problem, whether in central Wisconsin or in New Mexico. Valuable topsoil is lost, road ditches are filled, and small plants are cut off or abraded so that they fall prey to disease.

Eroded soil from uplands is carried by rushing water until the water slows down and drops its load of sediment in river channels, harbors, and reservoirs (Fig. 10.19). Great expenses for dredging are incurred. Depleted channels increase the threat of

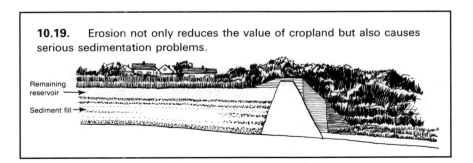

10.19. Erosion not only reduces the value of cropland but also causes serious sedimentation problems.

Remaining reservoir →

Sediment fill →

flooding, which causes untold hardships to homeowners and businesses. Dams built to control floodwaters and provide other services such as recreation and electric power lose their value as the capacity of their reservoirs decreases. In extreme cases, expensive dams have become a burden to the taxpayer in as little as 25 years.

It should be apparent that soil conservation pays. It pays where the soil is retained because the land can be more productive, and it also cuts costs downstream where the capacity of reservoirs is protected and water quality for domestic, irrigation, and recreational uses is preserved.

One benefit of sedimentation is that it results in the formation of alluvial soils in flood plains and at the mouth of rivers such as the delta of the Mississippi River.

Extent of the Problem

A natural resource inventory reported by the Natural Resources Conservation Service (NRCS), in 1992 revealed that the national average soil loss on all cropland due to water erosion is 3.1 tons per acre (6.9 Mg/ha [megagram/hectare]) per year. The figure varies for different parts of the country, ranging from 6.2 tons per acre (13.9 Mg/ha) per year in Alabama to 0.1 (0.2 Mg/ha) in Colorado. Many highly eroded areas experience losses in excess of 20 tons per acre (45 Mg/ha) per year. Drier regions with low amounts of water erosion may experience losses from wind erosion that match or exceed these values. Nationwide, wind erosion losses average 2.5 tons per acre (5.6 Mg/ha) per year. Comparative values for previous years were 3.2 (7.2 Mg/ha) in 1987 and 3.3 (7.4 Mg/ha) in 1982. NRCS reported the 1992 wind erosion figures for the individual states. The highest wind erosion occurred in New Mexico, which lost 13.5 tons per acre (30.2 Mg/ha) per year, while several states in the higher rainfall areas showed no wind erosion losses. Some of these states are Alabama, Arkansas, Connecticut, Georgia, Hawaii, and Kentucky. Some interesting comparisons are presented in Figure 10.20.

To place these figures in perspective, it is worthwhile noting that 1 inch (2.54 cm) of soil weighs about 150 tons (136 Mg). By using some of the soil-loss figures cited above, it is easy to calculate how long it takes to lose an inch of topsoil at those rates. For example, at 5 tons per acre (11.2 Mg/ha) per year, 1 inch (2.54 cm) of soil is lost every 30 years.

10.20.   Soil losses by erosion are typically great.

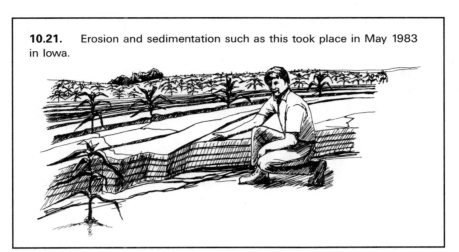

10.21. Erosion and sedimentation such as this took place in May 1983 in Iowa.

It is important to remember that eroded soil is not necessarily removed from a farm and transported to a lake or to the sea. It is more likely to be shifted from the high ground to the low ground, as illustrated in Figure 10.21. Eventually, the high ground becomes much less productive, and crops on the low ground may be damaged by sediment or those soils may become more productive.

Despite the great work of researchers and government agencies, the NRCS in particular, only about one-half the farmers in the United States have requested conservation plans for their land, and only half of those have carried them out. Some of the best-managed farms may have changed ownership and resulted in the soil conservation practices being abandoned by the new owners.

An excellent USDA publication by W. C. Lowdermilk in 1953, "Conquest of the Land through 7,000 Years," cites many examples of Old World states whose foundations were based upon agriculture and whose demise came when their abused land was no longer fruitful. In some cases, population increases placed too high a demand on the fragile soil, and in others political instability led to neglect of both the soil and the irrigation systems.

It may appear to some farmers today that soil conservation really is not very important. For instance, a yield of 150 bushels of corn per acre (9400 kg/ha) can be achieved on the black prairie soils of Illinois under an annual rainfall of 40 inches (1,000 mm). Wherever soil deteriorates by erosion (and that is just about everywhere in cultivated fields) and by compaction, which reduces soil aeration, farmers can still obtain high yields by adding more fertilizer, by planting more vigorous hybrid corn varieties, and even by irrigating in seasons of low rainfall. Thus good crop yields are achieved decade after decade, while the soil is slowly but surely wasting away unnoticed. News reports of crop surpluses in the United States and many other countries lull the public and politicians into complacency on the state of the soil resource.

CHAPTER **11**

Soil Classification

Scientific soil classification is generally recognized to have begun in 1885 when the Russian scientist V.V. Dokuchaev undertook the study and classification of soils near Moscow. At about the same time, in the United States, similar concepts of soils as natural bodies in the landscape were being formulated by E.W. Hilgard. C.F. Marbut, the director of the U. S. Soil Survey Division for the first 35 years of the 20th century, introduced many of the Russian concepts of soil science. Marbut's successor was C. E. Kellogg, who was the primary author of the first U.S. system of soil classification published in the U.S. Department of Agriculture (USDA) *1938 Yearbook of Agriculture*. It soon became clear that this system was inadequate, in part because it did not incorporate specific boundaries on soil properties in order for soils to be classified within a certain group. As a result, the soil survey staff began work on a modern system of soil classification in the 1950s. The product of their investigations and deliberations, under the leadership of Guy Smith, went through successive approximations and in 1960 *Soil Classification, 7th Approximation* was published and presented to participants in the 7th International Congress of Soil Science held in Madison, Wisconsin. After much worldwide scrutiny and many amendments, the 1975 edition of Soil Taxonomy was published. (Taxonomy is the science of classification.) At that time, 10 orders were recognized. The refinements continued, and in 1998 the current revision of *Soil Taxonomy*, with 12 soil orders, became available. It is a large volume with 869 pages.

Soil classification makes safe and productive uses possible for each kind of soil. In this time of increasing pressure on the land, the systematic approach of modern soil classification is a great help in avoiding abuse of soils and mistaken investments in land and operations that are incompatible with soil conditions. Soil classification and mapping (surveys) permit the transfer of soils information from one place to another and from the present to future generations.

The specific subdivision of soil science that deals with soil classification is *pedology*, and those who specialize in soil classification are pedologists.

Pedons and Polypedons

Soil horizons were described in Chapter 2 as the layers that form in the soil during the long period of soil development. When viewed in a two-dimensional cross section, such as the side of a pit, they represent a *soil profile*. A soil profile extends from the ground surface to the depth of soil development.

The concept of the soil pedon considers the soil profile in three dimensions. The *pedon* is described as the smallest three-dimensional body of soil large enough to illustrate the nature and arrangement of soil horizons and their variability. The surface area of a pedon is arbitrarily set at from 1 to 10 square meters, depending on the soil's uniformity or complexity. It is like a column of soil that would be left standing if a bulldozer removed all the soil except for that beneath a small patch of ground. Sometimes that is done if excavation takes place before the telephone company can reroute their lines.

A *polypedon* is defined as a set of contiguous pedons falling within the accepted range of characteristics for a specifically named soil on the landscape. The polypedon may also be termed a *soil body*. It is like having many columns of soil, side-by-side, extending to a boundary where a different kind of soil is encountered. The soil profile, pedon, and polypedon are illustrated in Figure 11.1.

Diagnostic Soil Horizons

Many types of soil horizons have been described by characteristics that fall within quantifiable physical and/or chemical parameters and meet a specified minimum thickness. At the highest level of soil classification, the soil order, there are three diagnostic surface horizons that are germane to the classification of most of the soils over the face of the earth. These named surface horizons may also be called *epipedons*. One of the epipedons, the mollic, has three named variations that will be discussed below.

Similarly, there are four diagnostic subsurface horizons used in the classification of most soils, at the order level, unless subsurface soil development is minimal. There is no word equivalent to "epipedon" for subsurface horizons.

Various combinations of these seven diagnostic surface and subsurface horizons determine most soil orders. There are three major exceptions. The first is where extreme climate, very dry or very cold, makes traditional agriculture impractical. The second is where the soil parent material is volcanic ash. The third is where there is a thick accumulation of plant debris, such as in bogs and tidal flats.

Seven common diagnostic horizons for soil orders are as follows:

Diagnostic surface horizons (epipedons)
    Ochric (pale or thin topsoil)
    Mollic (thick dark topsoil, neutral to alkaline, fertile)
    Histic (thick organic mat over mineral subsoil)

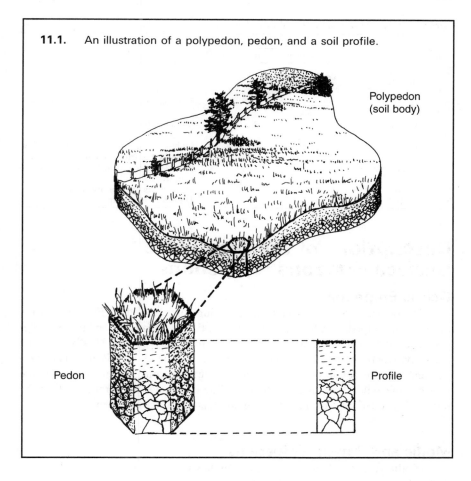

11.1. An illustration of a polypedon, pedon, and a soil profile.

Polypedon
(soil body)

Pedon

Profile

Diagnostic subsurface horizons

Cambic (only moderate soil development evident)

Argillic (enriched in clay washed down from above)

Spodic (enriched in colloidal humus, aluminum, and usually iron leached from surface horizons)

Oxic (severely weathered, infertile, high in sesquioxide clays, usually reddish colored)

A few other diagnostic horizons that are variations of those listed above will be discussed with the major diagnostic horizons.

In most cases, the epipedons have an A horizon symbol, whereas the diagnostic subsurface horizons normally carry a B horizon symbol. Lower case letters are used to indicate more specifically the type of soil development that has occurred. Table 11.1 explains the use of symbols in soil horizon nomenclature.

Table 11.1. Examples of soil horizons

Soil layer	Symbols General	Detailed	Properties
Solum			
Organic layer	O		
Leaf layer		Oi	Plant fiber recognizable
Humus		Oa	Saprophytes have decomposed the fibers
Topsoil	A		
Humus enriched		Ap	Darkened plowed layer
Subsoil			
Leached layer	E		Light colored due to fine particles being washed downward, eluviated
Accumulation zone	B	Bt	Where clay (German = tone) has moved in from above
Parent material	C		Little change by soil formation
Bedrock	R		The solid substratum

Descriptions of the Diagnostic Surface Horizons (Epipedons)

Ochric Epipedon

The ochric (from Greek *ochrose*, meaning pale) epipedon (Fig. 11.2) is the most common type of A horizon. It may be a thin A horizon or one that is either pale or dark colored if it has < 1.0% organic matter by weight. This is the usual condition where the native vegetation is a forest or the climate is arid. The average temperature conditions may be either warm or cold. When these soils are plowed, the fields have a grayish- or yellowish-brown appearance unless the farmer has added a lot of plant residue and manure to darken them.

Mollic and Similar Epipedons

Mollic (Latin *mollis*, soft) epipedons have a thick A horizon that is very dark brown or nearly black due to an enrichment of humus to > 1.0% by weight (Fig. 11.3). This condition is usually met where the native vegetation is prairie grass. The grasses cycle basic ions to the surface, and the limited precipitation prevents rapid leaching and maintains a high base saturation. These soils develop a strong granular structure, which allows them to be crumbly even when they are dry.

Three variations of the mollic epipedon are as follows:

Umbric epipedons appear to be mollic, but they have a low base saturation due to their acidity. The name comes from the Latin word *umbra*, meaning shade, which alludes to their dark color. Umbric epipedons are not widespread, but humid or wet conditions prevail where they occur.

Melanic (Greek *melasanos*, black) **epipedons** are black, humus-enriched A horizons formed in loose volcanic materials. They have a low bulk density, are high in aluminum, and have a high phosphate retention capacity.

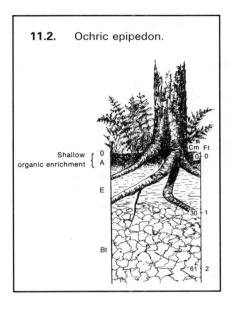

11.2. Ochric epipedon.

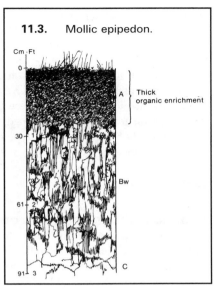

11.3. Mollic epipedon.

Anthropic (Greek *Anthropoes*, human) **epipedons** are a human-induced form of the mollic epipedon. They most frequently occur in arid regions that have a long history of irrigated agriculture with organic matter incorporation. They are formed extensively in the Orient.

Histic Epipedon

A histic (Greek *histose*, tissue) epipedon is an O horizon made up of plant residue 8 to 24 inches (20 to 60 cm) thick over mineral soil (Fig. 11.4). The range in thickness allowed depends on the degree of decomposition. These epipedons develop in lowlands that are saturated > 30 days per year. Histic epipedons are subdivided according to their degree of decomposition. From the least to the greatest, the terms used are **fibric**, **hemic**, and **sapric**. Their notations are Oi, Oe, and Oa, respectively.

Folistic (Latin *folia*, leaf) **epipedons** are O horizons that are similar to histic epipedons, but they can be as thin as 6 inches (15 cm) and are saturated for < 30 days per year. They form in upland positions and are common in alpine regions where low temperatures cause plant residue decomposition to be slow.

Descriptions of the Diagnostic Subsurface Horizons

Cambic Subsurface Horizon

When field and laboratory investigations of the subsoil show only a modest amount of weathering and not much accumulation of materials leached from above, the horizon is said to be cambic and has the symbol Bw. The word is derived from the Latin *cambiare*, meaning to change. To qualify as cambic, the horizon must not be very sandy and must show some alteration by processes of weathering. These changes may be evidenced by changes in color, the development of soil structure, or the removal of some of its more soluble components.

In the subhumid parts of the Great Plains, soils usually have a cambic horizon below the mollic epipedon where prismatic structure has developed and from which carbonates have been leached. Figure 11.3 shows a Bw horizon of this kind. Figure 11.5 shows another type of cambic horizon where periodic wetness has brought about a mottled color due to the form and distribution of iron. It is indicated by the symbol Bg. The "g" is derived from the Russian word *glei*, meaning wet sticky clay, but the "w" was used for no specific reason.

Argillic Subsurface Horizon

Throughout most of the humid hardwood forest region of the United States and some of the drier areas as well, the subsoils contain more clay than the A horizon and usually more than the C horizon. Some of the accumulated clay was moved down from the A and E horizons, and some was formed within the B horizon by the alteration of primary minerals into clay minerals. The small letter "t" in the horizon symbol Bt in Figure 11.6 is taken from the German word *tone*, meaning clay. This kind of diagnostic subsurface horizon is called an argillic

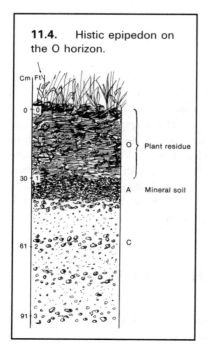

11.4.   Histic epipedon on the O horizon.

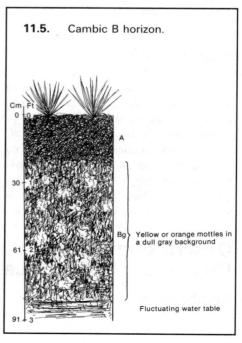

11.5.   Cambic B horizon.

horizon. The word is derived from the Latin *argillus*, meaning white clay. An argillic horizon usually benefits plants by holding moisture and nutrients within the root zone.

There are two special types of argillic horizons. One is the **natric (Btn) horizon**, which contains abundant sodium (*natrium* in Latin) that causes the soil to seal itself against the percolation of water. The impervious horizon illustrated in Fig. 5.20 is a natric horizon. The other is the **kandic (Bto) horizon** found in subtropical and warmer regions. They have very low-activity clay, primarily kaolinite, and therefore do not hold nutrients well.

Spodic Subsurface Horizon

In boreal (northern) forest regions and some wet sandy areas of subtropical regions, the subsoil is usually reddish brown to black. This color is caused by coatings of humus together with iron and aluminum oxides on the surfaces of sand grains. These coatings may break off and become tiny pellets within the soil matrix. A subsoil layer with these properties is called a spodic horizon (Fig. 11.7). It may carry the symbol Bh, Bs, or Bhs; h = humus and s = iron and aluminum oxides (sesquioxides). When these soils are plowed, the light gray overlying albic E horizon associated with the spodic horizon gives the field an ashy appearance. The name comes from the Greek *spodos*, meaning wood ash.

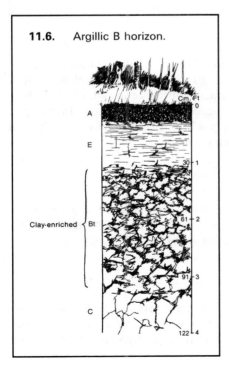

11.6. Argillic B horizon.

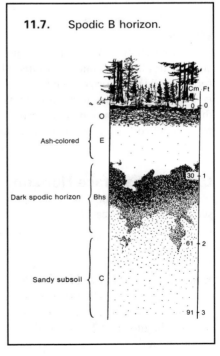

11.7. Spodic B horizon.

Oxic Subsurface Horizon

A very impoverished subsoil with almost no primary minerals other than quartz is called an oxic horizon (Fig. 11.8). It consists of quartz sand and an inert clay fraction of kaolinite plus oxides of iron and aluminum, hence, the symbol Bo. The entire subsoil is commonly quite uniformly weathered and lacks original rock features. Oxic horizons are found in tropical regions where severe weathering has been in progress for a very long time.

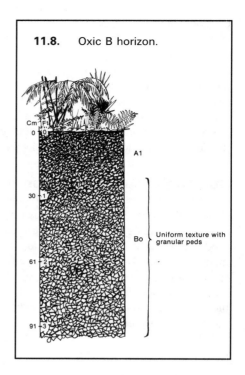

11.8.   Oxic B horizon.

Other Diagnostic Subsurface Horizons

Albic Subsurface Horizon

Albic (Latin *albus*, white) E horizons (Fig. 11.7) are whitish or gray in color with bleached, uncoated mineral grains. Usually they are below a mat of forest residue (O horizon) and at the surface of the mineral soil. The bleaching is, at least in part, due to organic acids leached from the O horizon. In this case, the albic horizon may be diagnostic for a soil order. Some may be shallow in the wet subsoil due to chemically reducing conditions.

Calcic Subsurface Horizon

A calcic horizon (Bk) (German *kalk*, lime) is an illuvial horizon in which secondary calcium carbonate or other carbonates have accumulated to a significant extent. They are widespread throughout grassland and desert regions.

Salic Subsurface Horizon

Salic horizons (Bz) (German *zalt*, salt) are layers with a high accumulation of soluble salts, typically sodium chloride, in arid regions, where they may be diagnostic. The salt is derived from periodically shallow saline water in the subsoil.

Hardpans as Diagnostic Horizons

In many parts of the world, a subsoil hardpan exists. The layers have a very important effect on the potential use of the soil. They are not typically diagnostic at the order level, but they are recognized as diagnostic for lower categories in the soil classification system. Four prominent ones are described in this section. An example of a horizon notation symbol is given for each.

Petrocalcic (Greek *petra*, rock) **horizons** (Bkm) occur in the subsoil on old landforms in arid

11.9. Petrocalcic B horizon.

regions where calcic horizon development has progressed to the point of becoming rocklike. It is composed mostly of calcium carbonate hardened around silicious gravel (Fig. 11.9). Petrocalcic horizons may extend to a depth of several feet. The "m" indicates induration.

Duripans (Latin *durus*, hard) are more durable than petrocalcic horizons because the cement in duripans includes much secondary silica (SiO_2). The symbol for them is Bq (q = quartz). They form best where there is or has been volcanic ash and the climate has alternating dry and wet seasons.

Fragipans (Latin *fragilus*, brittle) that occur in some forested regions are so dense they restrict the penetration of water and roots. The close packing of grains of sand and silt and weak cementation cause the horizon to be brittle when dry or moist, but not when wet. Most are Bx or Btx horizons.

Plinthite (Greek *plintos*, brick) horizons (Bv) form in warm humid regions where iron is abundant in parent materials. The iron becomes concentrated and cemented into a continuous reticulate (netlike) layer. Plinthite may harden irreversibly into an **iron pan** (petroferric layer) when exposed repeatedly to wetting and drying over a long time.

Soil Moisture and Temperature Regimes

In the classification of soils, *Soil Taxonomy* takes into account not only soil pedon characteristics, but also soil moisture and temperature regimes. Moisture regimes relate to the water available to plants in the main part of the root zone. Temperature regimes are measured at a depth of 20 inches (50 cm) or to a restrictive layer if it is shallower. Each regime has established parameters, but in this book only general features will be presented. The terms below are incorporated into the names of many of the soil taxonomic units.

Classes of soil moisture regimes

> *Aquic*–Saturated for enough of the time most years to cause reducing conditions (lack of oxygen) to prevail.
>
> *Aridic* and *torric*–Both terms are used to indicate dryness that restricts crop production without irrigation.
>
> *Udic*–Usually moist.
>
> *Ustic*–Seasonal dry periods, but enough precipitation during the growing season most years for crop production without irrigation.
>
> *Xeric*–Called a Mediterranean climate with dry summers and cool, moist winters.

Classes of soil temperature regimes

> *Cryic*–Very cold soils. Within this regime, the coldest soils have permafrost.
>
> *Frigid*–Cold winters, but summers are warm enough for crop production. Northern United States is an example.
>
> *Mesic*–Warmer than frigid. In the United States, the Ohio River Valley is an example.
>
> *Thermic*–Warmer than mesic. In the United States, the southern states are an example.
>
> *Hyperthermic*–Warmest of the temperate zone soils. Found in the hottest parts of the continental United States.
>
> *Isohyperthermic*–Hot tropical climate throughout the year.
>
> (The prefix *iso* can be used with most temperature regimes if the soil temperature is quite uniform throughout the year.)

The Soil Classification System Categories

The soil classification system of the USDA is hierarchical with these six categories:

Order
> Suborder
>> Great Group
>>> Subgroup
>>>> Family
>>>>> Series

The highest category, soil order, is the most generalized. As an individual soil is classified down through the system to the lowest category, soil series, an increasing number of specific properties is recognized. This system attempts to precisely categorize soils over the entire face of the earth in one of the 12 soil orders. It should be recognized that there are extensive areas of "not soil." These include rocky land, shifting sand, and ice/glaciers.

The 12 Soil Orders

In this book, the soil orders are grouped according to natural characteristics based on the five soil-forming factors in an effort to make them easier to remember. Each soil order has a formative element consisting of two to three letters that are underlined. These letters form the last syllable of the names in lower categories in that order. After each soil order heading, the estimated percentage of the world's soil in that soil order is given.

A world map of the 12 soil orders based upon *Soil Taxonomy* is presented as Figure 11.10.

A. Time Is too Short for Strong Soil Development.

Entisols: Soils Having Minimal Development (16.2%)

The order Entisol (English *recent*) includes soils that are so weakly developed they may have only an ochric epipedon and a nondiagnostic C horizon (Fig. 11.11). There are two reasons why these soils lack greater development: (1) The parent material consists of such highly resistant minerals that the rate of weathering is very slow. For example, droughty sands remain poorly developed because they contain an abundance of quartz. (2) The exposed land surface is young as a result of erosion or burial under new material brought in by wind or some other agent. Unprotected soils on slopes are subject to water erosion, and on more-level plains wind may erode the topsoil, thus exposing new material below and keeping the soil young. The opposite process keeps alluvial and wetland soils young. New material is added layer by layer as floodwater moves over the soils or temporary ponds spread across them.

Inceptisols: Immature Soils (9.8%)

Inceptisols (Latin *inceptum*, beginning) show more development than Entisols, but compared to other soils in the same region, they are immature (Fig. 11.12). They are found in most climatic zones but are excluded from arid regions and where there is permafrost. Most Inceptisols have an ochric epipedon and a cambic B horizon.

Sloping mountainsides are commonly occupied by Inceptisols because geologic erosion and leaching by rainfall is ineffective at such sites. Inceptisols are also common in depressions. One reason is that soil development is slowed where the soil does not dry out periodically.

B. Climate Is the Dominant Factor in Soil Development.

Aridisols: Desert Soils (12.0%)

Aridisols (Latin *aridus*, dry) are dry nearly all the year (Fig. 11.13). They have an ochric epipedon and may have either a cambic or an argillic subsurface

World soil map.

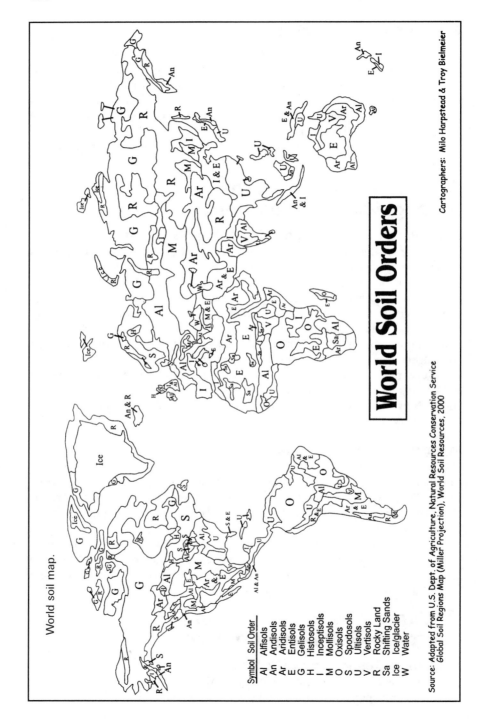

World Soil Orders

Symbol	Soil Order
Al	Alfisols
An	Andisols
Ar	Aridisols
E	Entisols
G	Gelisols
H	Histosols
I	Inceptisols
M	Mollisols
O	Oxisols
S	Spodosols
U	Ultisols
V	Vertisols
R	Rocky Land
Sa	Shifting Sands
Ice	Ice/glacier
W	Water

Cartographers: Milo Harpstead & Troy Bielmeier

Source: Adapted from U.S. Dept. of Agriculture, Natural Resources Conservation Service
Global Soil Regions Map (Miller Projection), World Soil Resources, 2000

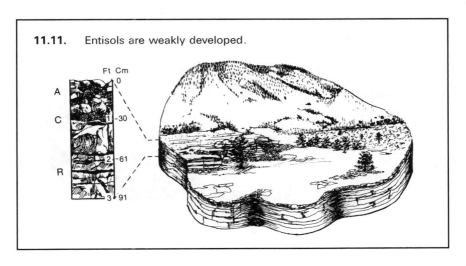

11.11. Entisols are weakly developed.

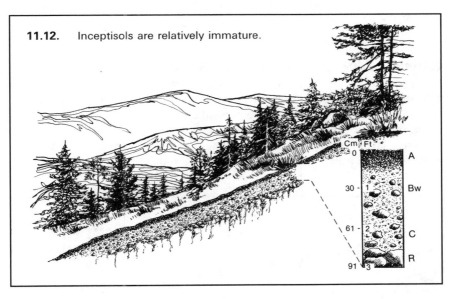

11.12. Inceptisols are relatively immature.

(B) horizon. Very young, gravelly sands, mostly Entisols, are common associates. Some, but not all, desert soils are salty. Most Aridisols contain lime, and in some the abundant lime has cemented a subsoil horizon, making a hardpan (petrocalcic horizon). Even more rocklike hardpans (duripans) have formed in places where there was cementation of soil with silica.

The deserts are very fragile regions, and once they are disturbed they are slow to recover. For this reason, intensive recreational use of the desert is viewed with alarm by environmentalists. Early records indicate that many Aridisol re-

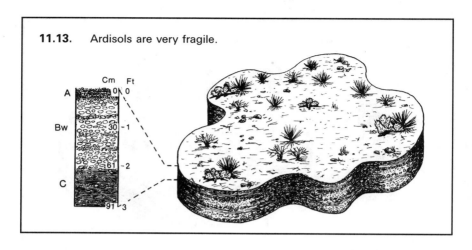

11.13.   Ardisols are very fragile.

gions were once quite grassy where now only scattered shrubs grow as a result of overgrazing and other uses.

Gelisols: Always Frozen Soils (8.6%)

The central concept of Gelisols (Latin *gelare*, freeze) is that they contain gelic material underlain by permafrost within 40 inches (100 cm) of the soil surface (Fig. 11.14). Gelic materials are mineral and/or organic matter that has been mixed in various patterns due to the churning caused by freezing and thawing in the active layer above the permafrost. They usually also exhibit ice segregation in this layer. Gelisols are found extensively in Alaska, Canada, and Siberia.

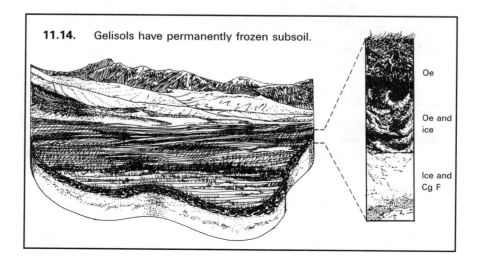

11.14.   Gelisols have permanently frozen subsoil.

<u>Ox</u>isols: Highly Weathered Tropical Soils (7.5%)

Oxisols (French *oxide*) form most commonly from sedimentary rocks and basic crystalline rocks that are relatively susceptible to weathering. An ochric epipedon overlies an oxic subsurface diagnostic horizon. Oxisols develop in tropical areas (Fig 11.15) and have a high content of inert clays, mostly amorphous oxides of iron and aluminum. The only kind of layered silicate clay found in more than trace amounts is kaolinite. Ironstone nodules, which may contain considerable amounts of manganese, are sometimes present in the soil profile.

Oxisols usually have a granular structure throughout, which allows them to absorb water readily and makes them easy to till. Nutrients are quickly lost from Oxisols when they are tilled because their humus decomposes quickly and their clays have a very low cation exchange capacity. Historically, farming has been carried out by a system of shifting cultivation, wherein the land is left to grow up to trees and shrubs for several years. This allows natural incorporation of nutrients from deep within the soil into the organic residue at or near the surface. Clearing, burning, and a few years of cropping follows while the humus decomposes. (This is called *slash and burn agriculture*.) When crop yields decline, the cycle is repeated. Many Oxisol areas are used successfully for permanent crops such as cocoa beans and oil palm. Some areas of Oxisols are planted regularly to sugarcane, pineapple, and other tropical crops with the help of modern agricultural techniques.

C. Parent Material Is Specific.
<u>And</u>isols: Volcanic Soils (0.8%)

Andisols (Japanese *ando*, dark soil) form in volcanic ejecta (pumice, cinders, lava) and closely associated parent materials in hilly or mountainous areas (Fig. 11.16). Fresh volcanic ash would not qualify as an Andisol and geologically

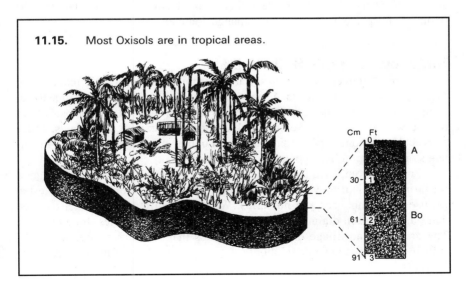

11.15. Most Oxisols are in tropical areas.

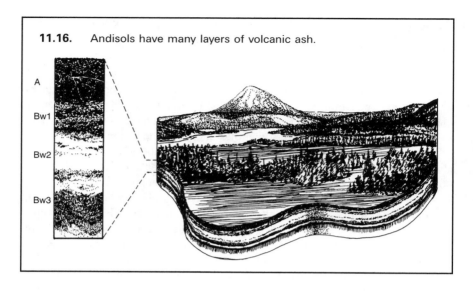

11.16.   Andisols have many layers of volcanic ash.

old deposits grade into other soil orders. Andisols have early weathering products in the colloidal fraction, namely allophane, imogolite, ferrihydrite, and aluminum-humus complexes. Due to the resistance to decomposition of the metal-humus complexes the organic matter commonly reaches 10% to 20%. This gives a melanic epipedon that has a very dark color.

Andisols have a low bulk density that promotes water infiltration, low water erosion potential, and ease of tillage. They are noted for their high natural fertility and are often the most productive soils in their region. Crops such as coffee are grown extensively on them. Nevertheless, production may be inhibited by the tie-up of phosphate on their anion exchange complex.

Histosols: Organic Soils (1.2%)

Histosols (Greek *histos*, tissue) are accumulations of organic matter (Fig. 11.17) in environments that have been too wet and/or cool for plant residue to decompose as fast as it has been produced by plant growth. Many Histosols have formed where a shallow lake or a tidal flat has filled in with dead plants. These soils act like a sponge and are saturated most of the time. Histosols on cool mountain slopes have a folistic epipedon and drain freely.

Poorly decomposed Histosols are commonly called ***peat*** and are not good for farming, even if they are drained. They are, however, sometimes harvested for use in greenhouses and nurseries. Well-decomposed Histosols are called *muck* and are often drained for specialized farming (vegetables, for example). The drainage and tillage of Histosols causes them to decompose, which results in subsidence at a rate of about a 1-foot (0.3-m) drop of the surface every 10 years.

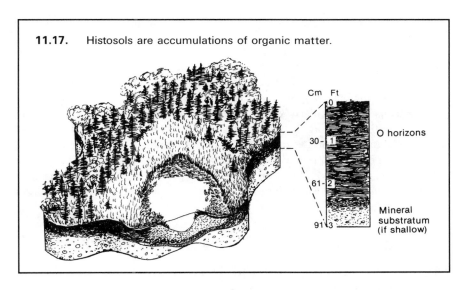

11.17. Histosols are accumulations of organic matter.

Vertisols: Cracking Dark Clay Soils (2.4%)

Vertisols (Latin *verto*, turn) are dark clay soils formed most extensively in the warm temperate and tropical areas with an ustic moisture regime, but they may also be found in cooler climates (Fig. 11.18). For example, in the United States, most Vertisols have been mapped in Texas, but they also occur in South Dakota.

Vertisols owe their unique properties to the shrinking and swelling of clays. Wide cracks open up during the dry season and some soil is likely to fall into them. When the rains return the cracks swell shut, but if they have been partially filled, there is not enough room for the cracks to close. This causes a churning ac-

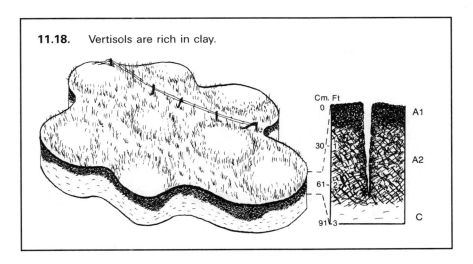

11.18. Vertisols are rich in clay.

tion that brings up fresh limy material from the C horizon and thereby rejuvenates the topsoil faster than it can be leached by rainwater.

The swelling action in Vertisols also causes lens-shaped blocks within the soil to slide past each other and develop polished surfaces called *slickensides.* The same forces buckle the landscape into mounds and hollows, which results in a microtopography called *gilgai.*

Vertisols are particularly well adapted to sugarcane and paddy rice culture. In Texas, Vertisols are largely used for pasture despite the fact that in dry seasons open cracks make footing hazardous for cattle. Within a few years, fence posts, telephone poles, and buildings may become tipped and twisted on these soils. When houses are built on them, the builder may provide a way to keep the soils under the foundation moist at all times to prevent the heaving action that can ruin a building.

D. Vegetation Is a Grassland (Prairie).

Mollisols: Grassland Soils (6.9%)

Mollisol (Latin *mollis*, soft) regions are among the most productive agricultural areas of the world (Fig. 11.19). The general properties of soils developed under prairie vegetation were discussed earlier in this chapter, and nowhere are they more strongly reflected than in the Mollisols. The dense fibrous root system of the grasses and forbs has resulted in the development of a thick, dark, humus-enriched A horizon (mollic epipedon) with an abundance of plant nutrients. These soils were first described in Russia, where the darkest were called Chernozems, meaning black earth.

In the humid part of the Mollisol area in the midcontinental United States—the corn belt—the subsoil has an accumulation of clay (argillic horizon). This property is minimal or even absent in drier parts of the grasslands where a cam-

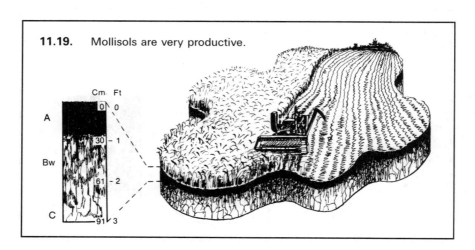

11.19. Mollisols are very productive.

bic diagnostic subsurface horizon prevails. In North America, Mollisols are most extensive on the treeless plains that extend eastward from the Rocky Mountains. Many settlers built their first houses of sod in which soil was bound together by grass roots.

E. Climate and Vegetation Combination Dominates.

Alfisols: High-base Status Soils of Hardwood Forests (9.7%)

Alfisols (from *pedalfer* in 1938 system) are typically found under deciduous forests where recycling of plant nutrients is effective (Fig. 11.20). Calcium, for example, is taken into the leaves and returned to the soil when they fall. Normally, the parent material contains calcium carbonate and is medium to fine textured. The humus-enriched A horizon is not thick and is therefore called an ochric epipedon. The subsoil has an accumulation of clay with dark clay films on the structural (ped) surfaces. This is the argillic horizon, which holds moisture and nutrients within the upper part of the root zone and is, therefore, beneficial to plants.

In the parts of the tropics where geologic erosion prevents the land surface from becoming highly weathered, the soils may grade from Inceptisols to Alfisols with the increasing stability of the landscape. When Alfisols are cleared of their timber and placed under cultivation they are usually quite productive and respond well to fertilization. The good supply of water, timber, and agricultural land in Alfisol regions throughout the world accounts for the development of large centers of population on them.

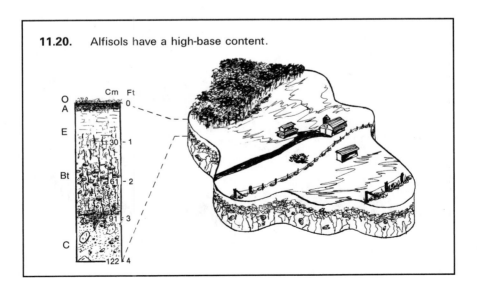

11.20. Alfisols have a high-base content.

Ultisols: Low-base Status Forest Soils of Warm Regions (8.5%)

Although cycling of bases (calcium, magnesium, potassium) goes on in Ultisols (Latin *ultimus*, last) under forest cover, it is less effective than in Alfisols because the geologic substratum usually lacks calcium carbonate. Leaching, which occurs year-round, has removed many plant nutrients from the root zone. There is a definite accumulation of clay in the subsoil (the argillic horizon), but it is highly weathered. In Ultisols the ultimate weathering of layered silicate clays has taken place. As a result, kaolinite is abundant. These soils are usually considered older by tens of thousands of years than the Alfisols. Well-drained Ultisols are brightly colored by stains of yellow and red iron oxides. Poorly drained Ultisols are gray.

If Ultisols are cultivated, they quickly become impoverished unless fertilization and careful management are practiced (Fig. 11.21). Historically, in the southern United States, many Ultisols were planted to cotton, which gave very low yields after a few years of production. This was followed by severe erosion and ultimate abandonment of the land, which continued to erode. By use of modern methods, many of these farms are being brought back into production; many even yield two crops per year due to the long growing season. In tropical areas three crops per year are possible with intensive management.

F. Vegetation and Parent Material Dominate.

Spodosols: Acid Soils of Sandy Pine Lands (2.7%)

Spodosols (Greek *spodos*, wood ash) are most common in the sandy outwash regions of the boreal forests and in quartzose (sandy) coastal marine de-

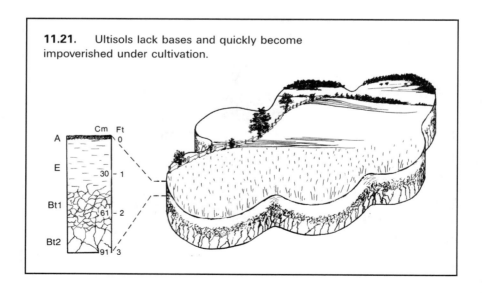

11.21.   Ultisols lack bases and quickly become impoverished under cultivation.

posits extending to the tropics. Under the acid humus layer (Fig. 11.22), there is likely to be a whitish albic horizon overlying a dark brown spodic horizon in which humus and/or iron oxides coat the sand grains. Sometimes these coatings cement this horizon into a pan called an ***ortstein.*** The Russian term *podzol* means ash beneath, in reference to the bleached layer under the forest litter. The village of White Earth, Minnesota, was named for this white soil horizon.

Much of the pine lumber that went to build towns and farmsteads throughout North America and Europe came from the extensive forests of the Spodosol regions. However, when the forests were cleared, these soils, which had yielded such beautiful timber, did not prove to be good for base-loving, shallow-rooted agricultural crops. Many farms failed and were replaced by pine plantations and mixed forests for lumber and pulp production as well as for use by wildlife and for recreation.

Lower Categories of the Soil Classification System

As was stated earlier, there are six categories in the USDA soil classification system. They are order, suborder, great group, subgroup, family, and series. Each of these categories is regulated by the Natural Resources Conservation Service of the USDA. The first five categories are defined in *Soil Taxonomy* and individual sheets are published for each soil series. In addition, two other categories, type and phase, may be described locally for land-use planning purposes. With each successive category, more information is revealed about the soil being classified.

In the following paragraphs, each category below soil order of the soil classification system is explained and examples are given.

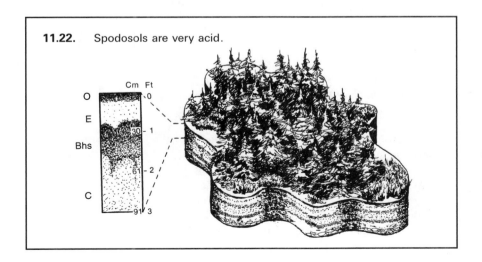

11.22. Spodosols are very acid.

Suborders

The suborder category uses a formative element (syllable) from the order name and places a new syllable before it to give more information about the soil. In many cases the new syllable may indicate such features as the usual moisture condition and a particular property of the parent material, or as in the case of Histisols, the degree of decomposition of organic materials. Some syllables may be used at more than one level in the soil classification system.

Five suborder syllables are given below with their meaning to illustrate how the system works:

> Ud—Latin *udus*, humid
> Psamm—Greek *psammos*, sand
> Orth—Greek *orthos*, true
> Cry—Greek *kryos*, cold
> Aqu—Latin *aqua*, water

By joining these syllables to selected formative elements from the order names, we get suborder names such as these (the common element is underlined):

> Ud<u>oll</u>—a M<u>oll</u>isol in a humid climate
> Psamm<u>ent</u>—a very sandy <u>Ent</u>isol
> Orth<u>od</u>—a true (most common) Sp<u>od</u>osol
> Cry<u>alf</u>—an <u>Alf</u>isol with a cryic or isofrigid temperature regime
> Aqu<u>ult</u>—a poorly drained <u>Ult</u>isol

Great Groups

Great groups are formulated by adding another syllable in front of the suborder name to give more information about the soil properties. Five examples of these are given below:

> Arg—Latin *argilla*, white clay
> Ust—Latin *ustus*, burnt (dry climate)
> Hapl—Greek *haplous*, simple
> Gloss—Greek *glossa*, tongue
> Pale—Latin *paleos*, old

Sometimes a vowel is placed between these syllables and the suborder name to make the word easier to say:

> Arg<u>iudoll</u>—a <u>Udoll</u> with clay accumulation in the B horizon
> Ust<u>ipsamment</u>—a <u>Psamment</u> of dry regions
> Hapl<u>orthod</u>—a simple <u>Orthod</u>
> Gloss<u>ocryalf</u>—a <u>Cryalf</u> with the E horizon tonguing into the B horizon
> Pale<u>aquult</u>—an <u>Aquult</u> showing evidence of great age

Subgroups

The subgroups are formed by modifying the great group name with one or two adjectives. This adjective may depict a normal condition, or it may indicate some special feature about a soil. In some cases a great group of one order may be integrated toward another, and this would be shown by the subgroup adjective. Five examples are given below:

Typic—the normal thing
Aquic—having properties of wetness
Alfic—grading toward an Alfisol (with an argillic horizon)
Fragic—having a fragipan
Aeric—periodic aeration

These terms may modify great groups to form subgroups as follows:

Typic Argiudoll—a typical Mollisol with an argillic horizon in a humid climatic zone
Aquic Ustipsamment—a slightly wet (seasonally) sandy Entisol in a dry climatic zone
Alfic Haplorthod—a simple, ordinary Spodosol having an argillic horizon below the spodic horizon
Fragic Glossocryalf—a cold Alfisol with tonguing and a fragipan
Aeric Paleaquult—an old, wet Aquult with colors indicating periodic aeration

The reader is reminded that each of these syllables indicates specific soil properties, as presented in the USDA publication, *Soil Taxonomy*.

Family and Series

The fifth category of the classification system is the *soil family*. This is not named with strange-sounding Greek and Latin syllables but instead has descriptive terms indicating such properties as particle size, mineralogy, cation exchange activity, and temperature regime. A common family name for many soils that are formed from glacial till in mid-America is coarse-loamy, mixed, superactive, mesic. A finer-textured, highly weathered soil in the southeastern United States might be in the fine-loamy, kaolinitic, subactive, thermic family.

The *soil series* is the sixth category and is the name given to soils with very similar profiles. The name that is given is derived from the town or community where the soil was first officially described. More than 20,000 soil series have been described in the United States and many more in other parts of the world. One example of a series is the Fayette, which is named after a town in Iowa. This soil formed in deep loess for some distance on each side of the Mississippi River in Iowa, Minnesota, Illinois, and Wisconsin.

Type and Phase

Soil type gives the texture of the tillage zone. *Soil phase* gives information about soil properties that affect land use; slope and stoniness are examples. The type and phase are not numbered here because they are not a part of the six-category system for classifying soils. However, type and phase are useful in land-use planning.

1. Order—Alfisol
2. Suborder—Udalf
3. Great group—Hapludalf
4. Subgroup—Typic Hapludalf
5. Family—Fine-silty, mixed, superactive, mesic
6. Series—Fayette
     Type—Fayette silt loam
     Phase—Fayette silt loam, nearly level

CHAPTER **12**

Soil Surveys

The classification of soils according to their distinctive properties was described in Chapter 11. When the various soils on the surface of the earth are delineated on maps, it is called a *soil survey.* Usually the mapping of soils is done on a county basis.

In the United States, the Natural Resources Conservation Service (NRCS) has the overall responsibility for making soil surveys to provide an inventory of our nation's soil resources. Agencies such as the U.S. Forest Service and the Bureau of Land Management also make soil surveys on lands for which they are specifically responsible. Through cooperative interagency efforts, the work by these agencies should blend in with that of the NRCS.

The soil scientists who make the soil maps are men and women who have graduated from a soil science program at an accredited university and completed a training period with an experienced soil surveyor. A knowledge of soil science for mapping purposes includes an understanding of geomorphology so that natural landforms can be identified. The boundaries of soil mapping units commonly coincide with the boundaries of various segments of a landscape, such as ridges, sideslopes, terraces, and floodplains. With sufficient experience in a locale, a soil surveyor should be able to predict the type of soil on any portion of that landscape and take soil cores just frequently enough to determine if the prediction was correct.

Figure 12.1 shows a small segment of a landscape *(left)* that has been expanded *(right)* to show its three component *soil bodies.* The Dubuque soil body has a silt loam surface horizon with clay subsoil over limestone bedrock. The Hixton soil body is a loam over yellowish-brown sandstone and siltstone. The Chaseburg is a deep silt loam formed in local alluvium. A farmer becomes familiar with the soil components of the landscape on his farm after years of tilling the soil and digging holes in it. The sequence of soil bodies down a hillside is called a *soil catena.* In the example shown, all soils are well drained, although the water table is closer to the surface the farther a soil body is downslope. In less-hilly terrain in humid regions, a typical catena may consist of dry soils at the ridgetop and wet soils at the footslope position.

12.1.   A soil landscape can be broken into several components.

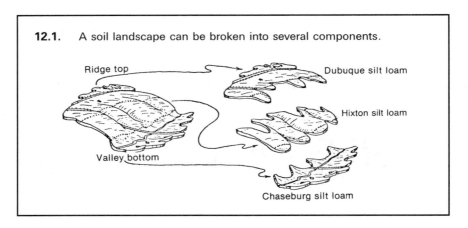

Landscape Patterns

The landscape is a mosaic of soil bodies that fit together as neatly as pieces in a jigsaw puzzle. Some landscapes have concentric, circular soil patterns; some are characterized by looped patterns; and others have parallel, linear soil patterns. The sketches in Figure 12.2 are simplified models based on a soil map, each representing 0.4 square mile (1 km²) of land in southeastern Wisconsin.

The patterns of landscapes influence farming, forestry, and wildlife management practices. For example, contour farming is not as easy on a landscape with tight circular patterns as it is in areas with linear soil patterns.

Making Soil Surveys

Since the 1930s, soil surveyors have been utilizing aerial photographs as a base map upon which boundaries between soil mapping units are drawn. The mapping units are determined by systematically traversing the land and augering up samples of soil, each of which is checked by sight and feel for its distinguishing physical properties and the slope is determined with a clinometer. Simple field tests may be conducted for pH, free lime, and soluble salts. The depth to which the soil is investigated will vary with its complexity of horizons, but it may

12.2.   Various soil bodies fit together to form the landscape.

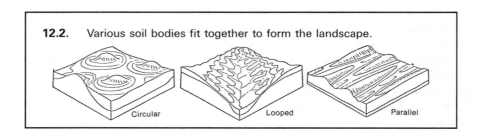

be as deep as 80 inches (2 m). Each mapping unit in the pattern of soils is given a symbol for the unit it represents (Fig. 12.3). The soil surveyor interprets all of this information, delineates the soil type on a base map (Fig. 12.4), and moves to another location.

Soil mapping is as much a matter of photo interpretation as it is soil investigation. The tone of black and white photos give a strong clue to the land form, degree of erosion of cultivated fields, the vigor of crops, areas of poor drainage, sometimes the species of natural vegetation, and much more to the trained eye.

Photo coverage for soil survey is made so that it provides stereoscopic coverage. This is accomplished by having a 60% overlap along the line of flight. When the photos are placed side by side, the soil surveyor can see photographed images for one position of the airplane with the left eye and another position with the right eye. This creates in the brain the appearance of a third dimension wherein objects with a higher elevation appear to rise up from the flat surface of the photos. In this way, the soil surveyor can properly record topographic features of the land.

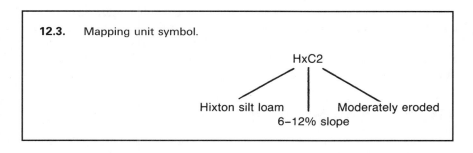

12.3. Mapping unit symbol.

HxC2

Hixton silt loam Moderately eroded
6–12% slope

12.4. A soil mapper makes auger holes to investigate the soil and records the findings on an aerial photograph.

Color photography has replaced much of the black-and-white film in recent years. In some places, color infrared photos of the land are available. They provide a better view of the pattern of vegetation and, by extension, the soil that favors a particular vegetative type.

The soil surveyor takes advantage of whatever geologic information is available. This is especially important when mapping soils where the parent material is residuum of bedrock as opposed to that which was transported by wind, water, or glaciers. In residuum, bedrock does not determine the type of soil that will develop, but it has a major influence on the soil properties.

The application of computers has provided soil surveyors with improved base maps by removing distortions inherent in photos made from an airplane. These corrected photos are called *ortho* (true) *photos.* When laid side by side they match. They can be digitized so that the photo image can be reproduced on a computer screen and zoomed in or out for a detailed or general view.

The global positioning system (GPS) was introduced in Chapter 8 and is becoming familiar to the general public. In the future, soil surveyors may carry a GPS locator and a compact computer loaded with a digitized base map that can be shown in three dimensions. In this way, more accurate information will be entered into a database as the fieldwork progresses. Experience has shown that such technical advances improve the quality and efficiency of soil surveying.

Uses of Soil Surveys

In the United States, the NRCS, agricultural universities, and county extension offices can usually supply information about the soils of a given county. For example, for Columbia County, Wisconsin, there is a 156-page published soil survey report with 122 map sheets showing soil mapping units in the county. Each unit has its own combination of soil horizons, slope, and moisture regime characteristics. As an example of the kind of information gathered, two of these soils are described in Table 12.1.

Just as the making of soil maps is changing, so is the publication of soil survey reports. Recent hard copy reports are more flexible, with only the photo section bound and the interpretive section in loose-leaf form for ease of modification. There is getting to be less emphasis on hard copy in the form of books containing photos, printed discussion, and interpretive tables. More emphasis is being placed on digitized photos and computerized interpretive information. This will extend the useful life of the reports because they will be easily updated and tailored to the specific needs of future users.

In the United States, soil surveys have been published for most of the agricultural regions as well as much of the forest and rangelands. In some cases, counties surveyed more that 30 years ago have to be rechecked to gather information that was not collected at the time of the original survey. The revised surveys are being digitized in accordance with the Soil Survey Geographic

Table   12.1.   Two soils of Columbia County Wisconsin

| Soil name | Kind of horizon | | | | Typical corn yield in bushels per acre[a] |
	Topsoil (A)	Subsoil (B)	Parent material (C)	Slope	
Plainfield loamy fine sand	Thick, pale (ochric)	None	Sand, acid to neutral	Undulating (2% to 6% gradients)	45[b]
Plano silt loam	Thick, dark (mollic)	Sticky to firm, blocky (argillic)	Silt over sandy loam glacial deposit	Nearly level (0% to 2% gradients)	130[c]

   [a] Without irrigation.

   [b] 45 bushels per acre = 2,800 kg/ha.

   [c] 130 bushels per acre = 8,160 kg/ha.

(SSURGO) database and must meet the national cartographic standards. As these efforts progress, more soils information will be added to what is currently available on compact disks and web sites.

Examples from the map section of a traditional soil survey report of Randall County Texas are shown here to illustrate some of the information available. Figure 12.5 shows a portion of a detailed soil survey map that might be used as the basis for making a farm plan. Figure 12.6 is a generalized map of the entire county suitable for making decisions about broad areas of crop and range management.

Soil surveys no longer benefit only agriculture but are of value to anyone who makes decisions about the land. This includes farmers and ranchers who want to maximize production efficiency. Fertilizer dealers better serve their customers through an understanding of the soils being managed. Engineers who bid on earth-modifying projects such as roads and airports find soil surveys useful for planning. Land developers must consider the soil for foundations, streets, lawns, and sometimes the septic systems. Bankers and other money-lending agencies can get a better feeling for the security of their loans if they know the potential of the land being used by the borrowers. Foresters use soil maps in species selection for regeneration, and they also plan the harvesting operations based, in part, on the bearing capacity of the soil and its susceptibility to erosion. Parks and other recreational facilities must be planned around the soil's suitability to support human and vehicular traffic. Tax assessors are becoming increasingly aware that they can make more equitable assessments on farmland and ranch land if they take advantage of the information provided in soil survey reports. As can be seen, there is a wealth of information available in these reports; only a few have been given here to illustrate the point.

Even if there is a published soil survey with a complete map of an area, it is a good idea to examine the soils. There is more variation on the landscape than the published map can show. It is helpful to learn how to recognize the properties of the specific soil horizons.

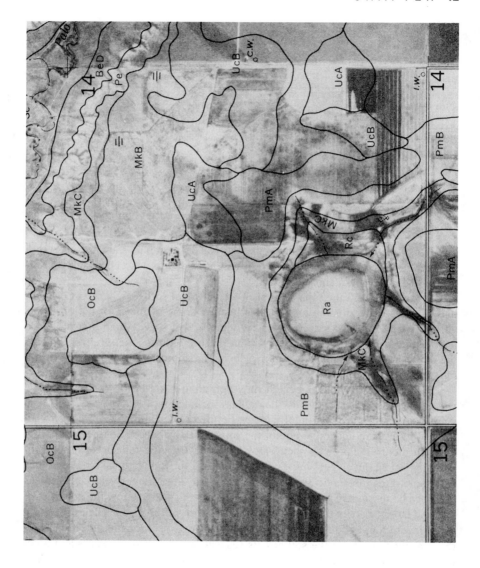

12.5. A detailed soil map of one section of land in Randall County, Texas. It is 1 mile (1.6 km) on each side.

Note: The percent slope designated by a letter, such as B, can vary with the individual survey, depending on the general steepness of the land in the county.

Legend: BeD = Berda loam, 5% to 12% slope; MkC = Mansker clay loam, 3% to 5% slope; OcB = Olton clay loam, 1% to 3% slope; Pe = steep rim of draws and canyons; PmB = Pullman clay loam, 1% to 3% slope; Ra = Randall clay of depressions or playas; Rc = Roscoe clay of depression edges; UcB = Ulysses clay loam, 1% to 3% slope.

12.6. A generalized map of Randall County, Texas.

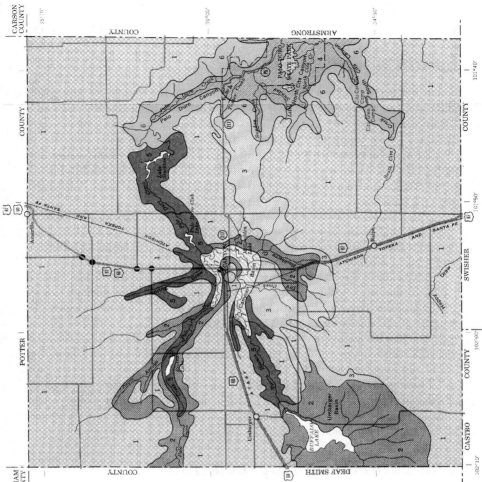

U. S. DEPARTMENT OF AGRICULTURE
SOIL CONSERVATION SERVICE
TEXAS AGRICULTURAL EXPERIMENT STATION

GENERAL SOIL MAP
RANDALL COUNTY, TEXAS

SCALE IN MILES

1 0 1 2 3 4

NOTE—
This map is intended for general planning.
Each delineation may contain soils having rat-
ings different from those shown on the map.
Use detailed soil maps for operational planning.

SOIL ASSOCIATIONS

1 Pullman association: Nearly level to gently sloping, deep soils that have a loamy surface layer and a firm clay subsoil

2 Ulysses–Mansker association: Nearly level to moderately sloping, loamy soils that are shallow and moderately deep over caliche

3 Olton–Amarillo association: Nearly level to moderately sloping, deep soils that have a loamy surface layer and subsoil

4 Rough broken land–Potter–Quinlan–Woodward association: Rough broken land and dissected, loamy soils that are very shallow to moderately deep over caliche, sandstone, or siltstone

5 Mansker–Berda–Potter association: Gently sloping to sloping, deep, loamy soils on foot slopes, and sloping to steep, loamy soils that are very shallow to moderately deep over caliche

6 Potter–Mobeetie association: Moderately sloping to steep, loamy soils on escarpments and foot slopes

7 Spur–Berda association: Nearly level to gently sloping, deep, loamy soils on flood plains and foot slopes

August 1968

191

Land Capability Classes

It has been indicated that soil survey maps are interpreted for many uses in the soil survey report, but in agriculture they are used most extensively to determine land capability classes. All soil bodies are placed in one of eight classes. Class 1 land is easily managed for crop production without having to overcome any appreciable limitations. Classes 2 through 4 have increasing limitations if they are to be tilled for crop production. Land in classes 5 through 8 is not recommended as crop land, and in class 8 the primary value is aesthetic and as a watershed.

The four possible subclasses—E = erosion, W = wetness, S = soil (e.g., shallow), and C = climate (e.g., arid)—show the dominant limitation that causes the soil to be placed in a particular class. The most obvious to cropping is the erosion hazard, which is based on slope.

Several drawings in other chapters can be used to illustrate some of the land capability classes. Class 1 land is depicted in Figure 3.9 because the land is level and no crop-limiting characteristics are shown. Figure 6.15 might be classed 2W because of its need for drainage. The erosion problem in Figure 10.2 is present because class 3E land is being improperly tilled up and down the hill. The shallow soil over bedrock in Figure 10.1B could be in class 6S, and the bog in Figure 11.17 might best be in class 7W.

Land has to be evaluated for several possible limitations to place it in the proper capability class, but by looking at these examples the reader should be able to understand better how the system works. Similar systems are used in the soil survey report for the land's potential for forestry and wildlife habitat.

Soil Landscape Appreciation

The eminent biologist Aldo Leopold wrote in his book, *A Sand County Almanac,* "When we see land as a community to which we belong, we may begin to use it with love and respect." It helps to know enough about the soil to evaluate soil landscapes intelligently and avoid mistakes in land management decisions. Landscape appreciation for its own sake is of interest to the ordinary citizen who does not have direct responsibility for soil management but feels an affinity to it.

CHAPTER 13

Soil Engineering

Soil is a source of material to be used in making a wide array of structures. Some examples include fill for dams and levees; foundation material for roads, runways, and buildings; aggregate (sand and gravel) for making concrete; clay for sealing the bottoms and sides of ponds, canals, and solid waste landfills; cover material over tanks, utility lines, tunnels, culverts, and conduits (sewers, drains, pipes for water, oil, and gas); and a porous medium for treating liquid wastes. Soil has been used to make homes like the adobes of the Southwest and the sod houses of pioneers settling on the American prairies. Soils support enormous loads, both inanimate (roads and buildings) and living (people, animals, and plants).

To the engineer, soil is any surficial material of the earth (or moon) that is unconsolidated enough to be dug with a spade. *Soil mechanics* is the field of engineering devoted to the use of soil as a building material. In the vocabulary of engineers, soil includes both the soil of the soil scientist and any loose substratum that may be present. The concepts of the soil as considered by the scientist and the engineer are merged in this chapter.

An advantage of using soil for engineering purposes is that there is so much of it, and it may be already on the site, which avoids the expense of hauling in other material. Another advantage is that soil can be so readily shaped into almost any desired form. Depending on how it is manipulated, soil can allow the passage of water through it or it can be made almost impermeable.

There are also some disadvantages for using soil in engineering. Soil is extremely variable, both geographically and over time. Cycles of wetting and drying as well as freezing and thawing change the engineering properties of soil. Unlike known types of steel or wood, soil is not a uniform material for which reliable strengths can be computed. Stable dry loam may be adjacent to unstable wet clay in a lowland. During a rainy period, the dry loam also may become wet and unstable. In winter both soils may freeze and heave in such a way as to crack pavements and basement walls, especially where the moisture content is high. Table 13.1 compares the suitability of three soils for various engineering uses.

Table   13.1.   Suitability or limitation rating for soils of the Clarion-Nicollet-Webster association. Reasons for limitations are given in parentheses

	Clarion	Nicollet	Webster
USDA classification	Fine-loamy, mixed, mesic Typic Hapludolls	Fine-loamy, mixed, mesic Aquic Hapludolls	Fine-loamy, mixed, mesic Typic Haplaquolls
Shrink-swell potential	Low	Moderate	Moderate
Roadfill	Good	Fair (wetness and low strength)	Fair (low strength, wetness, and shrink-swell)
Embankments, dikes, and levees	Severe (piping)	Moderate (piping)	Severe (wetness)
Dwellings with basements	Slight	Moderate (wetness)	Severe (wetness)
Septic tank absorption field	Slight	Severe (wetness)	Severe (wetness)
Sewage lagoon area	Moderate (slope and seepage)	Severe (wetness)	Severe (wetness)
Sanitary landfill area	Slight	Severe (wetness)	Severe (wetness)
Daily cover for landfill	Good	Fair (wetness)	Poor (wetness)

Source: Nelson, G. D. 1990. *Soil Survey of Murray County, Minnesota.* USDA-Soil Conservation Service, U.S. Government Printing Office, Washington, D.C.

Even though these soils are usually found next to each other in the landscape, their suitability for different engineering applications varies widely.

Engineering Properties and Classification of Soils

Preceding chapters have discussed soil physical properties from the perspective of factors relating to crop production. Soil mechanics also deals with physical properties as they relate to the use of soil as a building material. Some physical properties like particle size distribution and bulk density are important to both soil scientists and engineers. Even so, engineers have developed different soil classification systems specifically for engineering applications (Fig. 13.1). The Unified Soil Classification (USC) System was developed during World War II for the construction of military airfields and was later modified for use in foundation engineering. The American Association of State Highway and Transportation Officials (AASHTO) System is widely used by state transportation departments and the Federal Highway Administration for the design and construction of transportation lines.

Both the USC and AASHTO classification systems include several tests in addition to particle size distribution. Two standardized tests are often completed to test a soil's suitability as a building material. These tests are called the *Atterberg limits,* or the liquid and plastic limits. At a high water content, a soil possesses the properties of a liquid. As it dries, it acts more like a plastic, then like a semisolid, and finally like a solid when it is dry. The liquid and plastic limit tests are completed to identify the moisture content at which a soil changes from the consistency of a liquid to a plastic (liquid limit) and from a plastic to a semisolid (plastic limit). The Atterberg limits help engineers decide whether the soil material under consideration is suitable for their project or whether a different soil is needed. Liquid and plastic limit values are used with particle size information and other tests to classify soils in the USC and AASHTO systems.

Some of the soil characteristics that should be known before decisions about engineering uses are made are given in Table 13.2. Many soil survey reports include tables showing the suitability of the different soils for various engineer-

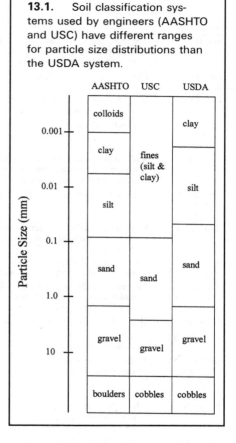

13.1. Soil classification systems used by engineers (AASHTO and USC) have different ranges for particle size distributions than the USDA system.

ing purposes. The American Society for Testing and Materials (ASTM) publishes official methods for measuring engineering properties of soils. It is important to understand the soils of an area by studying soil survey maps and reports, talking with soil scientists and engineers, and making the proper measurements.

Dams and Levees

Soil is the most readily available and least expensive material for the building of dams and levees. Numerous reservoirs, lakes, and ponds back up behind dams largely made of soil. Many cities located in floodplains are defended against overwhelming floods by miles (kilometers) of levees built of soil.

Such earthworks need to have two qualities: stability and impermeability. Four kinds of soil materials are needed to achieve these properties in a dam or levee: (1) a clay core and blanket are compacted and kept moist as a seal against

Table   13.2.   Characteristics of soils for engineering purposes

Kinds of information about soils	Comments
Soil texture	The inorganic (mineral) part of soils is a mixture of sand, silt, clay, and coarse fragments, even including boulders. The USDA, USC, and AASHTO systems are compared in Figure 13.1.
Kinds of clay	Clay species vary in degree of shrink-swell potential and other activity.
Depth to bedrock	Very shallow soils are usually unsuitable for excavation for basements, ditches along roads, or utility lines.
Kinds of surficial bedrock	Bedrock may be very hard (granite) or porous (sandstone, shale).
Soil density	Soil horizons range in density from porous to cemented. The denser soils have the higher bearing capacities and rate of transmission of vibrations (sound may travel twice as fast through dense soil as through air).
Content of rock fragments	Soil with many rock fragments is usually difficult to excavate and compact uniformly.
Erodibility	Many sandy soils are susceptible to wind erosion. Silty soils gully easily. Some clay soils are subject to piping, which is subsurface erosion by spontaneous tunneling.
Surface geology	The lay of the land affects land use. Proportions of steep and level land vary as well as soil pattern (linear, circular).
Soil pH (reaction)	Degree of acidity or alkalinity influences soil behavior physically, chemically, and biologically. To stabilize soil, engineers sometimes add hydrated lime to it, which raises the pH into the alkaline range.
Salinity	High salt content of soil affects its stability and that of vegetative cover.
Corrosivity	Soils differ in capacity to corrode buried pipes and tanks. Wet, acid soils are usually very corrosive.
Depth to seasonal water table	Soils with a seasonally high water table provide inadequate support to roads and structures. Frost action is most severe in wet soils.
Plasticity	Clayey soils are commonly quite plastic and, with wetting, become fluid like a thick liquid. The "plastic index" is the range of percent moisture content in which a soil is plastic.
Content of organic matter	To support growth of protective sod (Fig. 13.2), a soil layer containing organic matter is needed. Otherwise, organic soils, particularly peats and mucks, are usually removed in engineering projects.

leakage of water; (2) a sandy mass is added, surrounding the clay, to drain water away; (3) a layer of stone and rubble (riprap) is piled on the surface exposed to moving water such as waves and river currents; and (4) a good loam is used to cover remaining surfaces of the earthwork and to support the growth of protective vegetation (Fig. 13.2).

Sand is the most easily excavated and transported construction material because it is loose and does not become sticky upon wetting or hard upon drying. Bagged sand is used in emergency enlargement of levees during exceptional floods.

Because reservoirs and river floodplains tend to gradually fill with sediment washed in from upstream, this material often must be removed from the reservoir or channel (by dredging) every 20 to 50 years or so. The dredged sediment

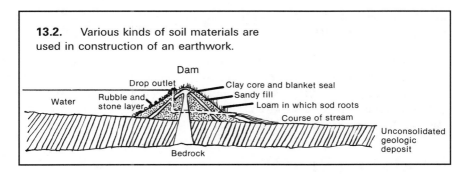

13.2. Various kinds of soil materials are used in construction of an earthwork.

can be used to build artificial islands or protected mounds on the floodplains. It can be returned to the farmland only at great expense. Reducing erosion is therefore doubly important for both soil conservation and reducing sediment buildup in waterbodies.

Roads and Buildings

Nearly all roads and small buildings are placed on soils, many of which are soft in wet seasons. In some areas subject to seasonal frosts, the soil heaves (lifts) or cracks due to desiccation. Dirt roads are impassable in certain seasons. Modern engineers find that naturally well-drained, very sandy, and gravelly soils provide the most trouble-free bases for roads and building foundations. On clay soils, a blanket of sand and gravel is placed over the clay before pavement is laid, and ditches are dug on either side to drain away storm and seepage waters (Fig. 13.3). Water, so essential for plant growth, is often undesirable in soils on which structures are placed. The long life of a road or building foundation depends on maintaining soil conditions that permit it to behave like a compact, well-drained sand or gravel.

Roadside soils also perform special functions unrelated to their engineering uses. Certain bacteria in soil convert poisonous carbon monoxide from auto-

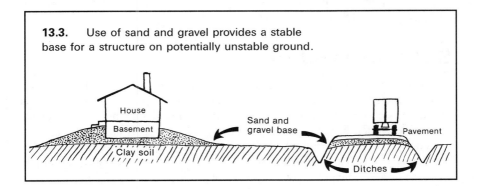

13.3. Use of sand and gravel provides a stable base for a structure on potentially unstable ground.

mobile exhaust into harmless carbon dioxide. Embankments of soil along major highways absorb much of the sound of traffic, thereby reducing noise pollution in the area.

The high capacity of sand and gravel to support weight arises from the particle-to-particle contact without lubrication of silt and clay between them. Gravel made of a strong and stable rock such as quartzite is long lasting under the stress of heavy traffic on pavement above the gravel bed. Concrete pavement itself contains as much as 50% gravel and 25% sand.

Ponds and Canals

Bottoms and sides of ponds and canals commonly need to be sealed to prevent leakage. Various kinds of lining materials may be used that are less expensive and less bulky than concrete. If clay is available locally, it may be mixed with bentonite, a special type of swelling clay, to make a tight clay liner. Fabric liners containing plastic or rubber or coated with asphalt may be used in the absence of clay.

Trails, Campgrounds, and Recreational Areas

Trafficability of soil under impact of human feet, horses hooves, and wheels of cycles and various other all-terrain vehicles deserves close attention. Nearly all soils are fragile with respect to traffic. Even sandy, gravelly loams cannot withstand trampling by horses for long without gullying or losing sand to the wind. Compaction of the soil can prevent plants from growing and can reduce infiltration, which leads to greater erosion by water. Trails, paths, and camping areas may be mulched (Fig. 13.4) with shredded bark or sawdust or may be covered with wooden walkways and platforms or asphalt pavement.

Rotation of trails, campgrounds, and play areas allows recovery of compacted soil and trampled vegetation during prescribed periods of nonuse. Soil degradation by heavy traffic can have long-lasting impact. Soil compaction by wagon traffic on the Red River and Oregon Trails in the mid-1800s is still evident today, some 150 years later.

Structures and Lines

Earth-sheltering of homes and burial of utility lines (Fig. 13.5) and tunnels protect structures and facilities from unfavorable temperatures that take place at or near the surface of the ground. However, some precautions should be followed.

Underground installations must be designed to support the great weight of soil cover. The strain on buried structures from the weight of overburden may be heightened in wet seasons by expansion of clays in the soil. Seepage of water into cracks in buried structures may be a recurrent problem. Growth of deposits of cal-

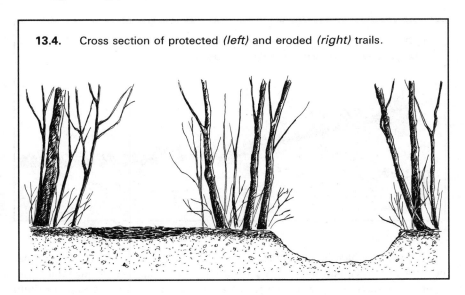

13.4. Cross section of protected *(left)* and eroded *(right)* trails.

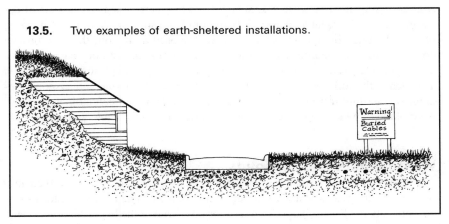

13.5. Two examples of earth-sheltered installations.

Warning!
Buried
Cables

cite or iron oxides in cracks may gradually shatter concrete belowground. Concrete and metal pipes and tanks are subject to corrosion in some moist soils (Fig. 13.6). Tiny electric circuits may develop spontaneously between the soil and iron pipes in such a manner as to literally bore minute holes in the pipes by dissolving the metal. In regions of permafrost, water mains must be insulated against freezing both from above and below.

Waste Treatment

Human activities generate liquid and solid wastes that require treatment and/or containment. Liquid wastes include residential and industrial wastewaters.

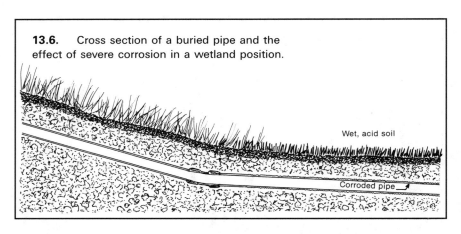

13.6. Cross section of a buried pipe and the effect of severe corrosion in a wetland position.

Wet, acid soil

Corroded pipe

Solid wastes include household and office wastes that consist of yard waste, paper, plastic, glass, and metal, and industrial wastes that include manufacturing byproducts such as sludges from paper mills. Because soil is a porous medium with enormous internal surface area populated by microorganisms capable of decomposing biodegradable materials, both liquid and solid wastes can be disposed of in soil. When biodegradable wastes are disposed of in soil, they are broken down and transformed mostly into water, carbon dioxide, and other gases. Nondecomposable wastes such as rock, metal, glass, plaster, and plastic remain buried or stored in the soil. Care must be taken to make sure that liquid and gaseous contaminants issuing from wastes decomposing in soils do not contaminate groundwater, lakes, and streams or come to the surface in unacceptable amounts.

Wastewaters and Biosolids

In urban areas, residential and industrial wastewaters are discharged into sewers and are treated at a wastewater treatment plant. The treated wastewater or effluent is typically discharged to a river or lake. One of the byproducts generated at wastewater treatment plants is sewage sludge, now called *biosolids.* Biosolids are typically over 99% water, contain the organic remains of treated wastewater, and are typically spread on agricultural land. Biosolids can be a source of plant nutrients and carbon for soil organic matter. Some biosolids from industrial wastewaters contain trace metals such as cadmium, chromium, lead, and zinc that contaminate soil and limit the use of biosolids on agricultural land. Dried biosolids (as low as 20% water content) are sold as a fertilizer. It can be a good soil amendment in the same manner as animal manure.

In rural areas, septic tanks and soil absorption systems or drainfields are used to treat residential wastewaters (Fig. 13.7). Wastewater is discharged from a house into a septic tank buried beneath the soil surface, where solids are removed and the wastewater undergoes primary treatment. The wastewater is then discharged to a

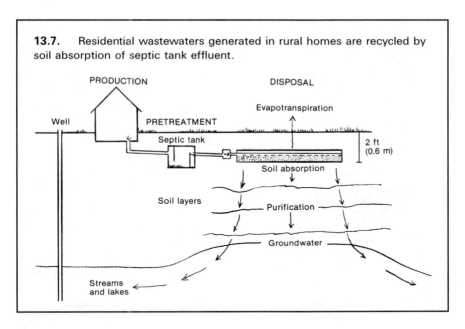

13.7. Residential wastewaters generated in rural homes are recycled by soil absorption of septic tank effluent.

gravel bed and the soil beneath the bed is used for final treatment and removal of contaminants. When soils contain too much clay or gravel or the depth to bedrock or groundwater is too shallow, an artificial soil mound of sandy or local soil material is constructed above the original soil surface. These types of systems remove some pathogens and nitrogen compounds from the wastewater.

Industrial wastewaters are the byproducts of manufacturing processes and vary in composition depending on the process. Some wastewaters, such as those generated at canneries, are often spray-irrigated on agricultural land after primary treatment (Fig. 13.8). Many other industrial wastewaters, if not discharged to wastewater treatment plants, are disposed of in stabilization ponds or absorption ponds. Stabilization ponds contain impermeable liners and rely on treatment within the pond and evaporation. Wastewater in absorption ponds is treated as it infiltrates through the soil beneath the ponds. Some industries generate hazardous wastes that must be treated and disposed of at special handling facilities. Researchers are currently developing alternatives to these methods of wastewater treatment, including discharge to constructed wetlands.

Solid Wastes

Solid wastes are typically buried in an engineered landfill that is made up of several cells that are filled in sequence over a period of several years. A compacted clay or synthetic liner at the bottom of each cell forms a seal to prevent infiltration of liquids that might contaminate the soil and groundwater beneath the

13.8.   Irrigation of farmland is being tried on a limited basis as a means of disposal for wastewater generated by some small industries such as canneries.

landfill. Liquid generated by the decomposition of solid waste is called leachate. Leachate is collected from beneath liners in pipes leading to a containment tank and then treated at a wastewater treatment plant. Solids placed in a cell are compacted daily, and when the cell is full, it is covered with a clay cap to minimize the infiltration of precipitation. Vents are installed in the landfill to allow gas to escape from slowly decomposing waste (Fig. 13.9). Soil and groundwater quality beneath and surrounding a landfill are intensively monitored to prevent environmental contamination. The contaminants of most concern are trace metals, nitrogen compounds, and organic solvents. Due to the large tracts of land needed and the potential for contamination, many municipalities are using alternatives to landfills, such as incineration, composting, and recycling.

Disturbed or Contaminated Lands

Disturbance of land, either by natural processes or human activity, and the contamination of land are often of concern to engineers. Disturbed and contaminated lands result in a situation where soil quality may be impacted, the growth of plants may be severely limited, or plants may not grow at all. Reclamation and remediation of a disturbed area is achieved by restoring it to a productive state. This may include artificial land forming and the

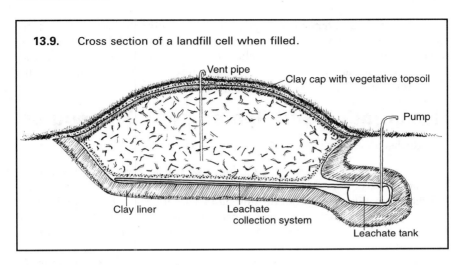

13.9. Cross section of a landfill cell when filled.

application of soil amendments and/or fertilizer to be followed by reestablishment of vegetative cover or enhancement of the soil environment to promote degradation of contaminants.

Naturally Disturbed Land

Soil may be disturbed or made unproductive by natural events, including landslides, floods that deposit sediment on lowlands, sand dune and dust invasions, blowdown of trees with consequent exposure of soil, and burial of land under fresh lava flows and volcanic ash falls.

Artificially Disturbed Land

Most disturbance of land is caused by human activities, including (1) mechanical strip mining for coal, oil shale, or metallic ore; (2) hydraulic mining of soil material for gold or phosphorus-bearing minerals used in fertilizer; (3) concentration of liquid and solid wastes (including mine tailings) on limited acreage; (4) contamination of soil areas with oil brine around oil wells and with toxic materials near chemical plants such as smelters and oil refineries; (5) contamination of soil beneath and surrounding leaking underground storage tanks; (6) sterilization with residues of agricultural chemicals in low spots in farmland; (7) quarrying; (8) construction (using cut-and-fill operations) in landscapes for development of residential and commercial buildings as well as roads and other facilities in urban areas; (9) operation of vehicles on fragile soils in deserts and tundra areas; (10) overgrazing of rangelands; and (11) overcultivation of croplands.

Many disturbed lands are left in the form of deep holes, mudholes, drifted sand, or mountains of overburden that are steeply sloping and where runoff is rapid. If this is the case, land forming is the first step in reclamation.

Disturbed soils are not necessarily lost to agriculture. At one site in Wisconsin, a sandy clay loam subsoil was excavated by a foundry company for making molds in which to pour and cool molten metal. The fertile loam topsoil was returned to the site, and it now produces high-quality crops just as it did before the excavation (Fig. 13.10).

Reclamation and Remediation Procedures

Reclamation may involve the following approaches:

1. *Confinement of objectionable substances.* Containment of wastewater and biosolids at wastewater treatment plants is necessary to prevent illegal and ecologically damaging spills from flowing into adjacent waters. Fertile agricultural soil itself becomes a polluting waste material if allowed to wash into streams and lakes. In this sense, the reason for preventing soil erosion is to confine the soil by keeping it in place.

Brine at oil wells is now commonly pumped back into deep layers in the ground, where it can be confined, instead of being poured out on the land.

Agricultural chemicals, especially pesticides, need to be confined aboveground, where they may be safely used or disposed of, to avoid contamination of the groundwater and the resulting pollution of drinking water.

As with biosolids, fly ash from coal-burning electric power plants, if free of heavy metal contaminants, may be spread on farmland as a fertilizer or soil conditioner. If heavy metal contamination is present, those substances must first be removed or the materials must be disposed of in an approved hazardous waste handling facility or landfill.

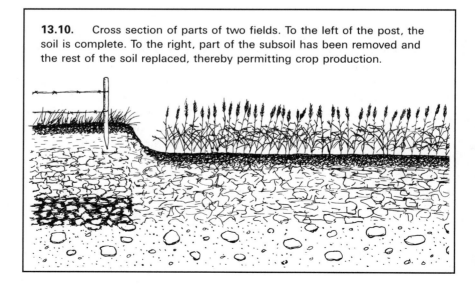

13.10.   Cross section of parts of two fields. To the left of the post, the soil is complete. To the right, part of the subsoil has been removed and the rest of the soil replaced, thereby permitting crop production.

2. *Removal of harmful substances from soil.* Excess salt in soil bodies can be leached downward if there is an adequate supply of freshwater and proper drainage underground or through conduits and ditches to dispose of the salty effluent. It is important that the soils be sufficiently permeable to allow the movement of water through them.

Where chunks of iron sulfide (pyrite and marcasite minerals) in mine tailings yield acid effluent, removal and proper disposal of the sulfide and the neutralization of acid with lime may be implemented.

When the soil beneath underground storage tanks has been contaminated with petroleum products such as gasoline, the soil surrounding the tank may be remediated by using several techniques. The soil is most often excavated and hauled away from the site and spread on agricultural land. Many of the contaminants evaporate or are degraded by soil microorganisms. At some sites, the soil is treated in place by enhancing the soil environment for degradation of contaminants (bioremediation). This technique involves injecting air into the soil, which drives some of the contaminants to the surface where they evaporate, while providing an environment more suitable for certain contaminant-degrading microorganisms. Researchers are currently trying to identify soil microorganisms that effectively degrade contaminants and could be introduced into an area where the soil is contaminated.

If the groundwater beneath these sites is contaminated, it also must be remediated.

3. *Land forming.* Shaping the land by construction of grassed diversion terraces and waterways can spread runoff and conduct water safely from sloping farmland. Where runoff is caught in holding ponds, the water can be used for irrigation or can simply be allowed to percolate to the water table. Properly maintained terraces can successfully subdivide long slopes into a sequence of short ones, thereby reducing both runoff and soil erosion.

Strip-mined land may be smoothed to slopes that are no steeper than the original ones. Correct stockpiling of topsoil and subsoil during the initial phase of mining makes it possible to restore the topsoil cover.

Some of the sediment that is dredged each year from streams, canals, and catch basins for the benefit of navigation and aquatic life may be useful on agricultural fields. Careful attention must be given to texture, organic matter content, pH, and other physical and chemical characteristics of the dredged materials (spoils).

Some abandoned quarries are filled with soil in such a way as to make them useful for cropping or other purposes.

In landscapes with a fairly high water table, standing water is used as a cover for disturbed land. Roadside excavations may become recreational lakes and ponds or sources of irrigation water.

4. *Interim protection of the soil surface during prolonged disturbance.* At some construction sites where disturbance of soil continues for months or even years, mulches of straw or sheets of special fabrics have been used to cover the bare soil until final cover of buildings, pavements, and lawns has been completed.

5. *Establishment of vegetative cover.* To reestablish vegetation, it is just as desirable to have a good seedbed prepared at the disturbed soil site as it is in a field to be planted to crops. In many cases, this may be difficult because of the nature of the soil material. In most cases, soil amendments and fertilizers may need to be applied. Disturbed soils may be high in some essential elements, but nitrogen is usually quite low.

Reestablishment of native species of vegetative cover by seeding and irrigation has been successful on land that was strip-mined for coal in the Four Corners area of the southwestern United States. In Ohio and Illinois, agricultural crops are being grown today on some prosperous farms that were inactivated for several years by strip-mining operations that ended with reconstruction of the landscape and its soils.

Researchers are currently trying to develop metal-scavenging plants called *hyperaccumulators* that remove metals from contaminated soil, especially in mined areas, and store them in their leaves and stems. The leaves and stems are harvested and the metals recycled.

Shelterbelts of trees and shrubs illustrate discontinuous establishment of vegetation for the protection of adjacent cropland from wind erosion. The trees reduce the wind speed, thereby reducing the amount of soil detachment and often creating a microclimate that reduces evaporation and increases yields.

Roadside pits from which construction materials have been removed may be stabilized by cut-and-fill operations and subsequent revegetation with trees and/or herbaceous cover.

To avoid the need for reclamation, every effort should be made to avoid contamination of the soil and to maintain it in acceptable form. To do this requires knowledge and dedication by those who use the land. Economic considerations have sometimes led to misuse of the land; thus, government regulations have been necessary to protect the rights of our citizens and to ensure productive and beautiful land for future generations.

GLOSSARY

AASHTO—American Association of State Highway and Transportation Officials. The AASHTO system of soil classification is used by engineers for the design and construction of transportation lines (roads, rail lines, and airport runways).

Accelerated erosion—Soil erosion increased by human activity beyond the normal or geological rate.

Acid—A substance with hydrogen ions available for chemical activity.

Acid rock—A rock such as granite that contains considerable amounts of silica and relatively little calcium, magnesium, and iron.

Acid soil—Soil with a pH value of less than 7.0, which is neutral.

Actinomycetes—Thread-like bacteria. Some fix atmospheric nitrogen symbiotically with nonlegume plants.

Aeration, soil—The process by which air in the soil is replaced by air in the atmosphere.

Aggregate—*See* Ped.

A horizon—The natural surface layer of mineral soil.

Albic horizon—A light-colored horizon just below the surface from which clays and humus have been leached.

Alfisol—An order of fertile deciduous forest soils that has an accumulation of clay in the B horizon.

Alkaline soil—A soil with a pH value of more than 7.0, which is neutral.

Alkali soil—A soil containing sufficient sodium to interfere with the growth of crops (same as sodic soils). The pH is normally 8.2 or higher.

Alluvial fan—Deposit from a stream as it enters a plain or larger stream.

Alluvial soil—A soil formed from alluvium (deposits made by rivers and streams).

Aluminosilicate minerals—Minerals composed largely of oxygen combined with silicon but with some of the silicon replaced by aluminum.

Amendment—Any substance added to soil that alters soil properties, such as gypsum, lime, fertilizer, and sawdust.

Ammonification—The release of ammonia by the microbial decomposition of protein.

Amphibole—A dark, basic mineral found primarily in granitic rocks.

Andisol—The order of soils formed from volcanic deposits.

Anion—A negatively charged ion.

Anion exchange capacity—The sum total of exchangeable anions that the soil can adsorb. Expressed as centimoles of charge per kilogram ($cmol_c$/kg) of soil material.

Anthropic horizon—A dark surface horizon enriched with organic matter due to human activity.

Apatite—A calcium phosphate mineral used as a source of phosphate for fertilizer.

Aquifer—An underground layer of permeable material that stores and can supply water.

Argillic horizon—B horizon of soil that contains more clay than the overlying A horizon. Some clay coatings are present on surfaces of blocky peds, having washed down from above.

Aridisol—A desert soil having a cambic or argillic B horizon.

Atmosphere—The layer of gas surrounding the earth: nearly 80% nitrogen, about 20% oxygen, and 0.03% carbon dioxide.

Atterberg limits—Liquid and plastic limits as measured by standard test procedures to determine a soil's suitability as a building material.

Available plant nutrient—*See* Nutrients.

Available soil water—*See* Soil water.

Azotobacter—Free-living bacteria that convert atmospheric nitrogen into organic nitrogen in the soil.

Basalt—A dark lava (igneous) extrusive rock that is high in iron, magnesium, and calcium.

Bases—Common parlance for ions of calcium, magnesium, and sodium in the soil.

Basic—Another word for alkaline, but often with emphasis on the presence of calcium and magnesium.

Basic rock—A rock with a high content of calcium, magnesium, and possibly iron and a relatively low content of silica.

Basin irrigation—Irrigation system often used in orchards where areas enclosed by ridges on all four sides are periodically flooded.

Bedding—Preparing a series of parallel ridges (beds) usually no wider than that of two crop rows, separated by shallow trenches usually less than the width between crop rows.

Bedrock—The solid rock underlying unconsolidated surface materials.

B horizon—Subsoil horizon lying below an A and/or an E horizon.

Biodegradable—Capable of being broken down into simpler products by living organisms.

Bioremediation—Any of several techniques for optimizing the physical, chemical, and biological conditions in the soil to promote the degradation and/or detoxification of pollutants.

Biosolids—A term for sewage sludge. A byproduct of wastewater treatment that contains solids having appreciable amounts of organic matter and nutrients but may also contain heavy metals and other potential contaminants.

Bog soil—A peat or muck such as a Histosol.

Border irrigation—Irrigation system where water is supplied to crops growing on gently sloping land between parallel ridges (borders).

Boreal—Northern, that is, cool.

Buffering capacity—Capacity of a soil to resist change, such as a change in the pH.

Bulk density—The mass of a dry soil sample per unit bulk volume (voids and all) as compared with the mass of an equal volume of water.

Calcareous soil—A soil containing enough free calcium (usually also magnesium) carbonate to show effervescence with acid.

Caliche—A layer near the surface, more or less cemented by secondary carbonates of calcium or magnesium precipitated from soil solution.

Calcic horizon—A subsurface horizon enriched with calcium and magnesium carbonates.

Cambic horizon—Weakly developed subsoil or B horizon.

Capillary action—The action by which the surface of a liquid is elevated or depressed depending on the relative attraction of the molecules of the liquid for each other or a solid with which it is in contact.

Capillary water—Water held in the very small pores of the soil.

Carbon-nitrogen ratio—The ratio of the weight of organic carbon to the weight of total nitrogen (mineral plus organic forms) in soil or organic material.

Carbon sequestration—The tying up of atmospheric CO_2 in plant tissue.

Carboxyl group—A grouping of carbon, hydrogen, and oxygen (COOH) that is present in organic acids in humus and other materials.

Catena—A "chain" or sequence of soils from the top of a hill to the footslope.

Cation—A positively charged ion.

Cation exchange—The exchange between a cation in solution with one on the surface of a soil particle.

Cation exchange capacity—The total of exchangeable cations that a soil can adsorb, commonly reported in centimoles of charge per kilogram (cmol$_c$/kg) of soil material.

Channelization—The deepening and/or straightening of natural drainage channels.

Chernozem—Black earth that is a Mollisol. Extensive in subhumid grasslands.

Chisel—To break up soil using narrow tools. Chiseling may be performed at other than the normal plowing depth. If done very deeply, it is usually called subsoiling.

Chlorosis—A lack of chlorophyll in a plant that results in a light green to yellow color of the plant tissue.

C horizon—Less developed soil usually below a B horizon.

Clay—Mineral material composed of particles less than 0.002 mm in diameter.

Coarse earth—The part of mineral soil that is too coarse to pass through a 2-mm sieve.

Colloidal particles—Clay and organic particles that are so small they tend to remain suspended in standing water.

Colluvial deposit—Soil or rock material gathered at the foot of a slope, primarily through the force of gravity.

Compost—Organic residues sometimes mixed with soil that have been piled, moistened, and allowed to undergo biological decomposition.

Conduction—Heat transfer due to the movement of kinetic energy between adjacent atoms in a substance brought about by a temperature gradient.

Convection—Heat transfer through the movement of a fluid (air or water).

Cover deposit—Loose material that blankets the surface of the substratum or bedrock. This includes loess, sand, gravel, glacial till and outwash, volcanic ash, alluvium, lake beds, shore deposits, and mass-wasting deposits.

Crust of soil—A somewhat dense, hard, or brittle soil layer at the land surface.

Crust of the earth—The outer 12-mile (19-km) thick layer of the lithosphere.

Crystalline rocks—Igneous and metamorphic rocks.

Cyanobacteria—Free-living bacteria that convert atmospheric nitrogen into organic nitrogen in crusts that were once called blue-green algae.

Dendritic pattern—A tree-like pattern that may be observed at erosion sites.

Denitrification—The microbial conversion of nitrate to the gaseous N form with subsequent release to the atmosphere.

Density—The mass (commonly expressed as weight) per unit volume of a substance.

Diagnostic soil horizons—The horizons that are the most important for soil classification.

Divalent cation—A cation having two positive charges, such as calcium (Ca^{++}).

Dolomite—Calcium-magnesium carbonate commonly called limestone.

Drainage—Water movement through soil. Natural drainage occurs when water drains out of the root zone to deeper layers or to groundwater. Surface channels or subsurface drains (tiles) can be used to artificially drain soils.

Drift—*See* Glacial drift.

Dryland farming—The practice of crop production in low-rainfall areas without irrigation.

Duripan—A resistant subsoil pan cemented chiefly with silica.

Eluviation—The removal of soil material in suspension (or solution) from a layer or layers of a soil.

Entisol—A very weakly developed soil.

Eolian (aeolian) deposit—Earthy materials deposited by wind. *See also* Loess.

Epipedon—A diagnostic surface soil horizon.

Erosion—The process by which soil is washed, blown, or otherwise moved by natural agents from one place on the landscape to another.

Escarpment—A steep face of the land at the edge of a region of high local relief.

Eukaryotes (eucaryote)—Organisms whose cells have nuclei. May be plant or animal.

Evaporation—The change of the state of water from a liquid to a gas, as happens at the surface of bare soil.

Evapotranspiration—The transfer of water vapor to the air by a combination of evaporation and transpiration.

Exchangeable cations—Cations such as those of calcium, magnesium, and potassium that are held loosely enough on surfaces of colloidal soil particles that they can exchange places with cations in the soil solution nearby.

Exchange complex—Surface of clay and humus having primarily negatively charged sites in most soils.

Fauna—The animals present at a site or in a region.

Feldspar—The most common primary mineral in the earth's crust.

Fertility, soil—The status of a soil with respect to its ability to supply the nutrients essential to plant growth.

Fertilizer—Any organic or inorganic material of natural or synthetic origin that is added to a soil to supply one or more elements essential for the growth of plants.

Field capacity—The percentage of water by weight that is held in the soil by capillary action after free drainage by the force of gravity has practically ceased.

Fine earth—That portion of a soil that is finer than 2 mm in diameter, including mineral sand, silt, and clay.

Flora—The plants present at a site or in a region.

Fragipan—Very dense subsoil layers that are brittle when moist and dry, but not when wet.

Furrow diking—Creating a dike and a small depression in furrows (normally 6 to 12 feet [1.8 to 3.6 m] apart) to hold water and reduce runoff.

Furrow irrigation—Irrigation system where water flows in furrows between crop rows.

Gabbro—A dark-colored crystalline igneous rock containing no quartz.

Gelisols—An order of soils characterized by having permafrost.

Genesis of soil—Soil formation.

Geomorphology—The study of landforms on the surface of the earth.

Gilgai—*See* Vertisol.

Glacial drift—Deposits made by glaciers and their meltwaters, including till and outwash.

Glacial till—Unsorted debris left by a glacier.

Glacier—A large body of slowly moving ice.

Gradient—A measure of soil in feet (meters) of rise or fall per 100 feet (31 m) of horizontal distance.

Global positioning system (GPS)—A global navigation system that uses satellites to triangulate positions on earth.

Gram-atomic weight—The atomic weight of an element in grams.

Granite—A light-colored, crystalline igneous rock containing considerable quartz (about 25%).

Green manure—A growing crop that is plowed under and mixed with the soil to enrich it with organic matter.

Groundwater—Water beneath the earth's surface in saturated soil or porous rock strata.

Gully erosion—Removal by water of sufficient soil to form channels large enough to prevent normal tillage from removing them.

Hardpan—A soil layer that acts as a barrier to the movement of water and the extension of plant roots.

Heat of fusion—The amount of heat necessary to change a substance from its solid to its liquid phase.

Heat of vaporization—The amount of heat necessary to change a substance from its liquid to its vapor phase.

Hectare—An area of land equal to 10,000 square meters or 2.47 acres.

Histic horizon—A largely surface soil horizon less than 18 inches (46 cm) thick that consists of about 25% by weight and 50% by volume of organic matter and is wet for prolonged periods.

Histisol—A soil order of peats and mucks that are thicker than 18 inches (46 cm). The soil material resembles that of the histic horizon.

Horizon—A natural soil layer, either surface soil or subsoil, that formed parallel to the land surface during the natural development of the soil body.

Humus—The dark, rather stable part of soil organic matter that remains after the major portion of animal and plant residues have decomposed and disappeared in the form of water and gases such as carbon dioxide.

Hydration—A mineral weathering process wherein water molecules combine with an oxide or a salt.

Hydrologic cycle—A cyclic pathway that describes water movement on earth. Includes such processes as evaporation, sublimation, precipitation, runoff, percolation, and groundwater flow.

Hydrolysis—A chemical reaction in which hydrogen and hydroxyl ions from water react with a mineral to form an acid and a base.

Hydrometer—A device used in determining soil texture based on the density of a suspension.

Hydrous mica—A common kind of layer silicate soil clay that is intermediate in responsiveness to wetting and drying between smectite and kaolinite clay.

Hyphae—Individual threads of mycelia.

Igneous rock—A rock that formed by the cooling and solidification of liquid parts of the lithosphere.

Illite—A hydrous mica type of silicate clay.

Illuviation—The process of deposition of soil material removed from one horizon to another in the soil; usually from an upper to a lower horizon in the soil profile.

Immobilization—The conversion of inorganic ions from the soil into organic molecules in living tissue.

Inceptisol—An order of moderately developed soils.

Infiltration—The downward entry of water into soil.

Infiltration rate—The rate at which water can enter the soil under specified conditions, including the presence of an excess of water.

Interveinal—Between the veins.

Ion—An electrically charged atom, with a surplus or deficiency of electrons.

Ironstone—*See* Plinthite.

Kaolinite—A 1:1 layer silicate clay with a low degree of responsiveness to wetting and drying and poor nutrient-holding capacity.

Landform—A natural feature of the earth's surface, such as a hill or a plain.

Latent heat—The amount of energy used to change the phase of a substance.

Laterite—Ironstone formed in soil by natural hardening of hydrous iron oxide with or without aluminum oxide and fragmental quartz. It is associated with plinthite and occurs in intertropical regions.

Lattice—A three-dimensional grid of lines connecting points that represent the centers of atoms or ions in a mineral clay.

Layer silicate clay—*See* Silicate clay.

Leaching—The removal of materials in solution by downward movement of water through soil.

Lime—Ground limestone, either calcite or dolomite.

Limestone—A rock rich in calcium carbonate derived from shells of sea organisms. Dolomite is often included in this term.

Limy soil—A soil containing appreciable amounts of carbonate minerals such as calcium or magnesium carbonate or a combination.

Lithosphere—The solid, rigid rock portion of the earth.

Loess—Deposit of windblown soil particles, largely silt size.

Macronutrient—A chemical element that is essential for plant growth and is used in relatively large quantities (usually >50 ppm in plants).

Magma—Liquid rock at great depths in the earth's crust.

Manure—Excreta of animals, with or without bedding material, normally added to soil to improve it with respect to crop production. *See also* Green manure.

Mass wasting—The movement downslope under the pull of gravity of large masses of soil and/or rock.

Matric force—The force of attraction between water and soil particles that holds capillary water in the soil.

Matrix—Something within which something else originates or develops.

Melanic horizon—A dark, organic-enriched surface horizon in some soils formed in volcanic deposits.

Metamorphic rock—Rock formed by recrystallization of igneous or sedimentary rock under great pressure and heat and by means of chemical reactions.

Micrometer—A standard measure of the length of 0.0001 mm. Micron (µm) is a synonym that is becoming obsolete.

Micronutrient—A chemical element necessary only in extremely small amounts (usually <50 parts per million in the plant) for the growth of plants.

Microrelief—Slight irregularities of a land surface.

Mineral—A natural inorganic compound with characteristic physical and chemical properties, commonly as an assemblage that forms rock.

Mineralization—The microbiological decomposition of organic matter into inorganic products.

Mites—Arthropods that are abundant in soil.

Mollic epipedon—Dark, thick, fertile topsoil best developed in grasslands.

Monovalent cation—A cation having a single positive charge, such as sodium (Na^+).

Montmorillonite—A very reactive 2:1 layer silicate clay. A member of the smectite group of clays.

Moraine—A blanket or ridge of unsorted debris left by a glacier.

Mottling—Spotted areas of color in a soil, usually associated with periodic wet conditions.

Muck soil—An organic soil that is more than half decomposed in the sense that visible organic fibers constitute a minor portion of the mass.

Mulch—A layer of material spread over the soil surface to protect it and plant roots from erosion, crusting, freezing, and drying.

Munsell colors—A system for uniformly describing the color of soil.

Mycelium—The thread-like mass hyphae of fungi.

Mycorrhizae—Fungi that live symbiotically with the roots of higher plants.

Myxomycetes—Slime molds; between bacteria and fungi.

Necrosis—Death or decay of plant or animal tissue.

Net radiation—The difference between incoming and outgoing shortwave and longwave radiation.

Nitrification—Biological oxidation of ammonium to nitrite and nitrate.

Nitrobacter—Bacteria that perform the second and final step of nitrification in the nitrogen cycle.

Nitrogen cycle—The biochemical changes that take place in repetitive sequence as organisms take up and release nitrogen.

Nitrogen fixation—Biological conversion of molecular dinitrogen (N_2) to organic combinations utilizable in biological processes.

Nitrosomonas—Bacteria that perform the first step of nitrification in the nitrogen cycle.

Nutrients—Substances essential for the growth of plants, such as nitrogen, phosphorus, and potassium. An available nutrient is one that can be readily taken up by the plant.

Ochric epipedon—Surface soil that is pale throughout or consists of a thin, dark layer over a pale layer of soil.

O horizon—A thin, surficial organic layer on soil, such as leaf litter.

Organic gardening—Gardening without the use of commercial fertilizers. Emphasis is on the use of composted materials, green manures, and judicious arrangement of plants and associated beneficial organisms.

Organic soil—A soil that contains a high percentage (12% to 18% depending on the clay content) of organic carbon throughout.

Ortstein—A natural soil pan formed in spodosols (Podzol soils) by cementation of the subsoil by organic matter and iron oxide.

Osmosis—The diffusion of water through a differentially permeable membrane, such as a root hair, from an area of high water concentration to an area of lower water concentration (or low salt concentration to high salt concentration) if pressure and temperature are equal on each side of the membrane.

Outwash—Deposit made by flowing meltwaters from glaciers.

Oxic soil horizon—A subsoil layer formed in intertropical regions.

Oxidation—A mineral weathering process wherein oxygen ions combine with multivalent elements such as iron.

Oxide clay—Fine particles composed of oxides of iron and aluminum, commonly in noncrystalline or amorphous forms.

Oxisol—An order of soils in which oxic horizons are prominent.

Parent material—Consolidated or unconsolidated material from which a soil forms.

Particle size distribution—Proportion of clay, silt, and sand particles in the fine earth of a soil.

Peat—A soil with relatively undecomposed fibrous organic material.

Ped—A unit of soil structure in a soil horizon; may be blocky, platy, or granular. A soil aggregate is the same as a soil ped.

Pedon—A volume of soil as it occurs in nature that is large enough to show the variations in the horizons. The surface area is declared to be from 1 to 10 square meters.

Percolation—The downward movement of water in soil.

Pesticide—Any chemical used to kill pests such as weeds, insects, and fungi.

Petrocalcic horizon—A highly developed and cemented calcic horizon. Often called *caliche*.

pH—The degree of acidity or alkalinity. The hydrogen potential expressed by a set of negative logarithmic values such that low numbers such as 3 and 4 signify great acidity and high numbers such as 8 and 9 signify alkalinity; pH 7.0 is neutral, that is, neither acid nor alkaline.

Plinthite—A nonindurated mixture of iron and aluminum oxides, commonly with some quartz and kaolinite clay, that has the capacity to harden into ironstone upon repeated wetting and drying.

Podzol—A great soil group, now replaced by the soil order Spodosol.

Pollution—The act of polluting, that is, to contaminate, make unclean or impure.

Polypedon (soil body)—Contiguous, similar soil pedons constituting a unit of land to the depth of soil development.

Porosity—The volume percentage of the total bulk not occupied by solid particles.

Prairie soil—A Mollisol formed under prairie vegetation.

Precipitation—Rainfall and snowfall plus minor amounts of dewfall and fog drip.

Primary mineral—A mineral formed naturally by crystallization from molten rock. Feldspar is the most common primary mineral in the earth's crust.

Productivity—The capacity of a soil to produce plant material. This is expressed in yield per unit of area per unit of time.

Profile—A two-dimensional, vertical cross section of a soil through all horizons.

Prokaryotes (procaryote)—Simple organisms whose cells lack nuclei. Not clearly plants or animals.

Quartz—Crystalline silica such as is present in granite. A resistant primary mineral.

Quartzite—A metamorphic rock consisting of quartz grains cemented with silica.

Radiation—Heat transfer by electromagnetic waves.
Reduction—A mineral weathering process in the absence of oxygen that causes multi-valent elements such as iron to become more soluble.
Regolith—Loose earth materials above solid bedrock.
Relief—The elevation, or differences in elevation considered collectively, of a land surface on a broad scale.
Rhizobia—Bacteria that convert atmospheric nitrogen into organic nitrogen in root nodules of legumes.
Rhizosphere—A zone of the soil where plant roots are in abundance and soil microbes are especially active.
Rhyolite—A fine-grained extrusive equivalent of granite.
Rill erosion—Water removal of soil in channels small enough to be filled in by normal tillage.
Runoff—Water from precipitation or snowmelt that flows over the soil to surface water bodies.
Salic horizon—A layer of soil enriched with soluble salts.
Saltation—The leaping, jumping, or bouncing of soil particles along the surface of the soil during wind erosion.
Sandstone—A sedimentary rock, usually of quartz, bound together by a cementing material such as silica or iron oxide.
Saprolite—A mass of weathered rock.
Saturation—The condition of a soil when all pores are filled with water. Base saturation percentage refers to the proportion of bases on the cation exchange material.
Secondary mineral—A mineral formed by weathering from a primary mineral. For example, kaolinite is a secondary mineral formed from feldspar. All carbonates such as calcite and dolomite are secondary minerals.
Sedimentary rock—A rock composed of sediment, that is, deposits made by water, wind, ice, and gravity.
Sensible heat—The amount of energy used to warm the air above the soil surface.
Shale—A sedimentary rock made of clay, silt, and very fine sand.
Sheet erosion—Water removal of a thin layer of soil from the soil surface.
Silicate clay—Clay composed of minerals made up largely of crystalline layers of silica and alumina.
Slickenside—Polished surfaces caused by blocks of soil sliding past each other during the formation of Vertisols.
Smectite—A group of very reactive silicate clays of which montmorillonite is a prominent member.
Sodium adsorption ratio (SAR)—The ratio of soluble sodium to soluble calcium and magnesium. SAR is used to assess the quality of irrigation water and the risk of soil dispersion.
Soil—The loose mass of broken and chemically weathered rock mixed with organic matter that forms on the earth's surface.
Soil absorption system—A system of pipes buried in the soil through which effluent from septic tanks infiltrates into the soil.
Soil body—A component or unit of the natural terrain (landscape).
Soil classification—A shorthand system to provide detailed soil description.
Soil formation (genesis)—The processes by which parent material is transformed into a body of soil.
Soil heat storage—The amount of energy stored as heat in the surface soil layers.
Soil landscape—The soil portion of the landscape.
Soil management—The sum total of all tillage operations, cropping practices, fertilizer, lime, and other treatments conducted on or applied to a soil for the production of plants.

Soil moisture regimes—A system for describing the predictable available moisture in the crop root zone of the soil throughout the year.

Soil reaction—Degree of acidity or alkalinity (pH).

Soil sequence—An array of soils (soil bodies) such as from the top of a hill to the foot-slope or from the youngest soil on the youngest landform surface to the oldest soil particles in the same region.

Soil solution—The aqueous liquid phase of the soil and its solutes.

Soil structure—The clustering of soil particles into units called peds or aggregates.

Soil survey—The identification, classification, mapping, scientific and practical explanation, evaluation, and interpretation of the soil cover of a terrain.

Soil temperature regimes—A system for describing the predictable soil temperature throughout the year. It has implications for crop production.

Soil texture—The relative proportions of the three soil separates (sand, silt, and clay).

Soil water—Water contained in soil. Available soil water is that which can be taken up by the plant.

Solum—The A and B horizons together or the single one of these that overlies the C horizon at a site.

Spodic horizon—B horizon in which sand grains are coated with aluminum oxide, humus, and usually iron oxide; best developed under coniferous forests.

Spodosol—An order of soil in which the spodic horizon is prominent.

Sprinkler irrigation—A system whereby water is applied to a crop by stationary or moving sprinklers.

Stomata—Minute openings in a leaf that permit gaseous interchange.

Stratification—Layering of water-laid deposits such as sand and gravel.

Stubble mulch—The stubble of crops or crop residues left essentially in place on the land as a surface cover before and during the preparation of the seedbed and possibly during the growing of a succeeding crop.

Subirrigation—Irrigation system where the groundwater table is adjusted so that adequate water is available to the crop rooting systems.

Sublimation—The change of state for water from a solid to a vapor. The transformation of snow or ice directly to water vapor.

Substratum—A layer beneath the surface soil.

Summer fallow—A management practice in some dry zones wherein during alternate cropping seasons vegetative growth is prevented by shallow cultivation in order to store soil water for the next year's crop.

Surface creep—The rolling by wind of coarse sand particles over the soil surface.

Suspension—The movement by wind of small soil particles (silt and clay) caught up by air currents.

Symbiotic nitrogen fixation—The close association of certain bacteria and legumes to convert atmospheric nitrogen (N_2) into nitrogen forms usable in biological processes.

Tectonic activity—Disruption of the earth's crust resulting in earthquakes, volcanoes, faults, and related events.

Terrace—A bench-like landform on valley bottoms ("high bottoms").

Thermal conductivity—Property of a substance that indicates its ability to conduct heat.

Tillage—Mechanical manipulation of soil for any purpose; in agriculture it is usually restricted to the modifying of soil conditions for crop production.

Tilth—The physical condition of soil as related to its ease of tillage and fitness as a seedbed.

Topography—The lay of the land, that is, levelness or hilliness.

Transpiration—The transfer of water to the atmosphere through the stomata of plant leaves.

Trickle irrigation—Also called drip irrigation. A system whereby small emitters drip water onto the soil adjacent to the plants, or the tubing and emitters are placed below the surface in the upper part of the root zone.

Ultisol—An order of infertile (relative to Alfisols) soils of warm region forestlands with clay accumulation in the subsoil.

Umbric epipedon—Dark, deep, surface soil horizon that is more acid than a mollic epipedon.

Unsaturation—The state of a soil in which most of the smaller pores are filled with water and the larger ones are primarily filled with air.

USC—Unified Soil Classification system. The USC system of soil classification is used by engineers involved in foundation engineering.

Vermiculite—A 2:1 layer silicate clay derived from hydrous mica.

Vertisol—An order of soils high in montmorillonite clay (smectite group) in the dark, thick, surface soil that tends to heave when wetted after a dry period, causing a wavy microtopography (gilgai).

Watershed—An area draining ultimately to a particular body of water.

Water table—The surface of the groundwater.

Weathering—The disintegration and decomposition of minerals and rock.

Wilting point—The condition of a soil when its water content is so low that plant roots can no longer obtain adequate water to sustain life.

Windbreak—A planting of vegetation to protect downwind crops from desiccation and breakage and to protect soil from wind erosion.

INDEX